AF905095

Translational Systems Sciences

Volume 11

Editors in Chief
Kyoichi Kijima, Tokyo, Japan
Hiroshi Deguchi, Yokohama, Japan

Editorial Board
Shingo Takahashi, Tokyo, Japan
Hajime Kita, Kyoto, Japan
Toshiyuki Kaneda, Nagoya, Japan
Akira Tokuyasu, Tokyo, Japan
Koichiro Hioki, Tottori, Japan
Yuji Aruka, Tokyo, Japan
Kenneth Bausch, Riverdale, GA, USA
Jim Spohrer, San Jose, CA, USA
Wolfgang Hofkirchner, Vienna, Austria
John Pourdehnad, Philadelphia, PA, USA
Mike C. Jackson, Hull, UK
Gary S. Metcalf, GA, USA
Marja Toivonen, Helsinki, Finland
Sachihiko Harashina, Ichikawa, Japan

In 1956, Kenneth Boulding explained the concept of General Systems Theory as a skeleton of science. He described hopes to develop a "spectrum of theories–a system of systems" which may perform the function of a "gestalt" in theoretical construction. Such "gestalts" in special fields have been of great value in directing research towards the gaps which they reveal".

There were, at that time, other important conceptual frameworks and theories, such as cybernetics. Additional theories and applications developed later, including synergetics, cognitive science, complex adaptive systems, and many others. Some focused on principles within specific domains of knowledge and others crossed areas of knowledge and practice, along the spectrum described by Boulding.

Also in 1956, the Society for General Systems Research (now the International Society for the Systems Sciences) was founded. One of the concerns of the founders, even then, was the state of the human condition, and what science could do about it.

The present Translational Systems Sciences book series aims at cultivating a new frontier of systems sciences for contributing to the need for practical applications that benefit people.

The concept of translational research originally comes from medical science for enhancing human health and well-being. Translational medical research is often labeled as "Bench to Bedside." It places emphasis on translating the findings in basic research (at bench) more quickly and efficiently into medical practice (at bedside). At the same time, needs and demands from practice drive the development of new and innovative ideas and concepts. In this tightly coupled process it is essential to remove barriers to multi-disciplinary collaboration.

The present series attempts to bridge and integrate basic research founded in systems concepts, logic, theories and models with systems practices and methodologies, into a process of systems research. Since both bench and bedside involve diverse stakeholder groups, including researchers, practitioners and users, translational systems science works to create common platforms for language to activate the "bench to bedside" cycle.

In order to create a resilient and sustainable society in the twenty-first century, we unquestionably need open social innovation through which we create new social values, and realize them in society by connecting diverse ideas and developing new solutions. We assume three types of social values, namely: (1) values relevant to social infrastructure such as safety, security, and amenity; (2) values created by innovation in business, economics, and management practices; and, (3) values necessary for community sustainability brought about by conflict resolution and consensus building.

The series will first approach these social values from a systems science perspective by drawing on a range of disciplines in trans-disciplinary and cross-cultural ways. They may include social systems theory, sociology, business administration, management information science, organization science, computational mathematical organization theory, economics, evolutionary economics, international political science, jurisprudence, policy science, socio-information studies, cognitive science, artificial intelligence, complex adaptive systems theory, philosophy of science, and other related disciplines. In addition, this series will promote translational systems science as a means of scientific research that facilitates the translation of findings from basic science to practical applications, and vice versa.

We believe that this book series should advance a new frontier in systems sciences by presenting theoretical and conceptual frameworks, as well as theories for design and application, for twenty-first-century socioeconomic systems in a translational and transdisciplinary context.

More information about this series at http://www.springer.com/series/11213

Anssi Smedlund • Arto Lindblom
Lasse Mitronen
Editors

Collaborative Value Co-creation in the Platform Economy

Springer

Editors
Anssi Smedlund
School of Business
Aalto University
Helsinki, Finland

Arto Lindblom
School of Business
Aalto University
Helsinki, Finland

Lasse Mitronen
School of Business
Aalto University
Helsinki, Finland

ISSN 2197-8832 ISSN 2197-8840 (electronic)
Translational Systems Sciences
ISBN 978-981-10-8955-8 ISBN 978-981-10-8956-5 (eBook)
https://doi.org/10.1007/978-981-10-8956-5

Library of Congress Control Number: 2018942016

© Springer Nature Singapore Pte Ltd. 2018
The chapters "Expanding the Platform: Smart Contracts as Boundary Resources" and "Distributed Governance in Multi-sided Platforms: A Conceptual Framework from Case: Bitcoin" are licensed under the terms of the Creative Commons Attribution 4.0 International License (http://creativecommons.org/licenses/by/4.0/). For further details see license information in the chapters.
This work is subject to copyright. All rights are reserved by the Publisher, whether the whole or part of the material is concerned, specifically the rights of translation, reprinting, reuse of illustrations, recitation, broadcasting, reproduction on microfilms or in any other physical way, and transmission or information storage and retrieval, electronic adaptation, computer software, or by similar or dissimilar methodology now known or hereafter developed.
The use of general descriptive names, registered names, trademarks, service marks, etc. in this publication does not imply, even in the absence of a specific statement, that such names are exempt from the relevant protective laws and regulations and therefore free for general use.
The publisher, the authors and the editors are safe to assume that the advice and information in this book are believed to be true and accurate at the date of publication. Neither the publisher nor the authors or the editors give a warranty, express or implied, with respect to the material contained herein or for any errors or omissions that may have been made. The publisher remains neutral with regard to jurisdictional claims in published maps and institutional affiliations.

Printed on acid-free paper

This Springer imprint is published by the registered company Springer Nature Singapore Pte Ltd.
The registered company address is: 152 Beach Road, #21-01/04 Gateway East, Singapore 189721, Singapore

Preface

The purpose of this book is to present a collection of both conceptually and empirically grounded fresh academic ideas combining two increasingly popular topics in management research – service science and platforms. The premise of the book is that value is co-created in purposefully created and managed service systems that in the era of digitalization are connected to larger wholes, systems of systems, showing evolutionary characteristics. Service systems are being increasingly brought together, managed, and orchestrated with platforms: modular technical or nontechnical, physical or virtual venues capable of acting as foundations on which a company in the market can build their business.

Combining the perspectives of platforms and value co-creation is still a nascent approach. Previously, platforms were regarded only as technological modular architectures. They were thought of as onion-like multilayered structures where the core is an IPR-protected technology. Along with the advent of platform business models and the so-called platform economy, the definition of platforms has broadened. At the same time, research into service dominant logic and value co-creation has resulted in a new and inclusive perspective on how value is created in relationships. Although these ideas are relevant for all economic activity, they are more pronounced, scaled, and multiply with advancements in connectivity and ICT.

Despite the booming interest, we are still at the beginning of platform economy research. The majority of platform research has not adopted the value co-creation ideas of the science of service systems. However, there are researchers and research groups who are conducting ambitious efforts in combining these two perspectives. The book in hand describes results of such efforts as well as results of others closely related to the service science community. We are publishing four fully conceptual chapters and nine chapters with empirical data or anecdotal case evidence. They cover a wide range of topics, including literature review on the state of the art in service platforms, service platform orchestration, smart contracts, social contracts, the effects of digitalization on society, public sector digital platforms, crypto currencies, retail, collaborative consumption, management of IoT in manufacturing, AirBnB, and regional innovation ecosystems.

We begin with four conceptual chapters. Chapter "Service in the Platform Context: A Review of the State of the Art and Future Research" by Xia Han, Veronica Martinez, and Andy Neely provides a systematic literature review on service in the platform context. Chapter "Platform Ecosystem Orchestration for Efficiency, Development, and Innovation", written by Anssi Smedlund, Hoda Faghankhani, Heini Ikävalko, and Petra Turkama, is about types of service platforms and how platform owners orchestrate for efficiency, development, and innovation. Joni Salminen, Nicolas Gach, and Valtteri Kaartemo, in chapter "Platform as a Social Contract: An Analytical Framework for Studying Social Dynamics in Online Platforms", conceptualize platforms in light of social contracts theory in order to provide understanding of the social processes in the platform ecosystems. Chapter "Expanding the Platform: Smart Contracts as Boundary Resources", a contribution from Kristian Lauslahti, Juri Mattila, Taneli Hulkkinen, and Timo Seppälä, is also about contracts, but those executed by computer programs in interactions between the platforms and their environment.

The next two chapters are overarching societal and multilevel takes on the platform economy. Antti Hautamäki and Kaisa Oksanen explore the challenges and possibilities of digital platforms for the public sector with the case example of Finland in chapter "Digital Platforms for Restructuring the Public Sector". William Rouse, Michael Pennock, Zhongyuan Yu, and Kara Pepe clarify how digitalization and platforms affect society, organizations, processes, and work practices and then elaborate on the topic with case examples in healthcare.

The next three chapters are about retail or consumption. In chapter "Understanding Platform Transformations Through Routine Interactions", Lauri Paavola examines Tesco – a UK-based retail giant from a routine interactions point of view. In chapter "Creating the Foundation for a Functioning Internal Platform", Mikko Hänninen, Olli Rusanen, and Lauri Paavola apply an internal platform construct to the evolution of the strategy of a Finnish retailer. Arto Lindblom and Taru Lindblom take a consumer perspective in chapter "Applying the Extended Theory of Planned Behavior to Predict Collaborative Consumption Intentions": they apply the theory of planned behavior to predict collaborative consumption intentions.

The last four chapters illustrate the multitude of contexts and emerging topics relevant for service platform research. Juri Mattila and Timo Seppälä in chapter "Distributed Governance in Multi-sided Platforms: A Conceptual Framework from Case: Bitcoin" explain how blockchain technology enables distributed governance among actors and what consequences it may have for competition. Chapter "Value Creation from the Internet of Things in Heavy Machinery: A Middle Manager Perspective" by Marko Sommarberg, Robin Gustafsson, Zeerim Cheung, and Eero Aalto provides a managerial perspective of a manufacturing firm adopting an Internet of Things (IoT) strategy. In chapter "Customer Value in the Sharing Economy Platform: The Airbnb Case", Hong Ngoc Nguyen, Timo Rintamäki, and Hannu Saarijärvi provide an extended framework of customer value and illustrate the framework with the case of AirBnB. Last but not least, Santi Novani, Cici Cintyawati, and Lidia Mayangsari provide an overlook on an Indonesian regional innovation ecosystem and its collaboration platforms in chapter "Back to the Future: A Revelation of Conventional Platform Preference of Digital Creative Ecosystem Entities in Bandung".

As guest editors, we believe that this present volume will promote growing interest in the topic of value co-creation in the platform economy and provide fruitful avenues for further research. As a part of the Translational Systems Science series, we hope that this volume promotes platform topics in the science of service systems.

Acknowledgments

As guest editors, we wish to express our gratitude to the reviewers who devoted their time to the manuscripts, providing valuable feedback to the authors. We especially would like to thank Professor Kyoichi Kijima for providing us the opportunity to create this volume. We also want to express our gratitude to the organizations that have funded our scientific activities: the Finnish Funding Agency for Technology and Innovation (TEKES) and, particularly, Global Marketplaces (GlobaMa) project that has enabled us to focus on the various forms of the platform economy in the retail industry. GlobaMa project was conducted at the Aalto University School of Business in 2016–2018 together with corporate partners Kesko Corporation, Posti Group, Unilever Finland Oy, and Solteq.

Contents

Service in the Platform Context: A Review of the State of the Art and Future Research

Xia Han, Veronica Martinez, and Andy Neely

Abstract Traditional ways of doing business have been turned upside down by a group of new companies. Uber is "the world's largest taxi-company owns no vehicles". Facebook is "the world's most popular media owner creates no content", and Airbnb is "the world's largest accommodation provider owns no real estate". This relatively new research phenomenon requires a comprehensive understanding. This systematic literature review explores and questions "platform" research in the context of services.

This article studies 133 articles between 2002 and 2016. The findings suggest that the service platform is an increasingly popular field of research with a wide spectrum of disciplines across 74 journals. It is gaining momentum moving from theoretical to an empirical research area. Ninety-one articles included empirical data.

The definition of a service platform has been categorized in three major groups. They are the architectural researches (n = 37), the economic group (n = 21), a comprehensive definition (n = 48) and generic (n = 27), which is a group of papers that did not explicitly discuss the core features of platforms. This literature review develops a taxonomy of research topics based on their research focus: (1) service architecture, (2) platform's impact on services and (3) service platform strategy. Finally, three key challenges are identified, which also serve as opportunities for future research.

Keywords Service platform · Multi-sided platform · Literature review

X. Han (✉) · V. Martinez · A. Neely
Cambridge Service Alliance, Institute for Manufacturing, University of Cambridge, Cambridge, UK
e-mail: xh268@cam.ac.uk; vm338@cam.ac.uk; adn1000@cam.ac.uk

© Springer Nature Singapore Pte Ltd. 2018

A. Smedlund et al. (eds.), *Collaborative Value Co-creation in the Platform Economy*, Translational Systems Sciences 11,
https://doi.org/10.1007/978-981-10-8956-5_1

Introduction

New technologies have enabled the proliferation of platform-based business models across industries, drastically changing the landscape of today's economy. Uber, Facebook and Airbnb can all be categorized as "multi-sided platforms" (hereafter referred to as "platforms"). These platforms serve the function of matching the needs and resources of two or more groups of customer (Evans and Schmalensee 2016; Parker, Van Alstyne and Choudary 2016). One of the most distinctive features of these platforms is the positive correlation between the number of participants and the value of the network (Hagiu and Wright 2015; Gawer 2009).

Despite the extraordinary impact of "platforms" in our service economy, the existing literature is mainly focused on product-based platforms (Thomas et al. 2014). Service accounts for over 50% of the GDP of the developed world's economy (World Bank 2014). Researchers are beginning to explore the "service" aspect of platforms (Suarez and Cusumano 2010; Gawer 2011); therefore, the service platform agenda is an open subject for future research. The objective of this chapter is to investigate the state of the art in terms of "service platforms". The systematic literature review was selected because of its strong objectivity and transparent approach to searching for and synthesizing research (Tranfield et al. 2003).

This chapter is structured as follows: first, the methodology used to select the relevant papers is briefly introduced. Then, the research findings, trends and future directions are discussed. Finally, the limitations and conclusions are presented.

Methods

This explorative review follows a six-stage process proposed by Tranfield et al. (2003): scope and identification of key words, evaluation of search results, refinement of search criteria, title and abstract review, selection of articles for full review and synthesis.

Scoping

Use of the term "platform" is very broad, varying from a concrete digital marketplace to a saloon facilitating discussions. This study takes a slightly narrower view. "Platform" in this study requires the article to contain explicit mentions or implications of network effects. For instance, in the information systems and information technology literature, the term "platform" has been loosely used as equivalent to "system" or "software". For instance, Tyagi and Senthil (2015) discuss the process of moving library automation software to a cloud-based platform. In this case, platform is dismissed, since the core service activity, library automation, does not

Fig. 1 Scope of the literature

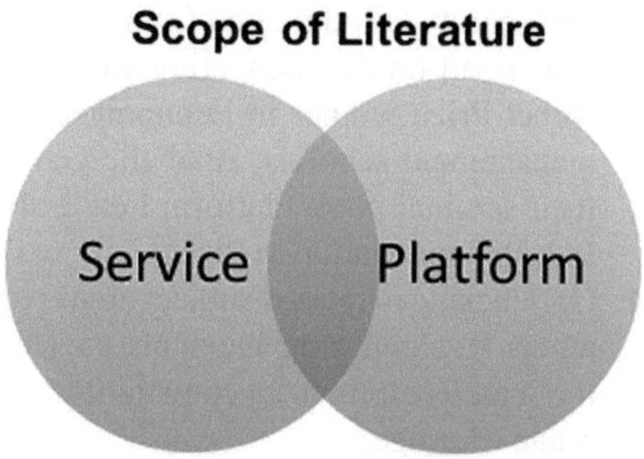

Table 1 Initial search result

Key word/databases	Science Direct	Web of Science	Google Scholar
Platform	58,015	313,702	3,960,000
Service	197,126	875,007	5,960,000

benefit from network effects; nor does the paper extensively discuss technical implications such as modularity.

Comprehensive Search

First, the key words in the study were identified. In order to capture the widest range of literature while remaining relevant and focused, two of the most representative key words were chosen, namely, “service” and “platform” (Fig. 1). Only singular forms of the key words were chosen because their plural forms are automatically searched for by the databases.

Three databases were selected to test the search terms, with each database representing a segment of database size. The basic search strings representing the entire knowledge base were tested across the three databases. Science Direct returned the lowest number of results, and Google Scholar generated the highest number of matches. Table 1 shows the initial search results.

All three databases generated a significant amount of literature. However, the majority of the findings were not in management-related fields and were irrelevant to this systematic literature review. Therefore, a set of exclusion criteria was set up to filter the research results. Through this process, the comprehensiveness of the databases in the relevant fields was further tested.

The following criteria were applied at the refinement stage:

1. Only English articles were chosen for the first two databases, where such options were available.
2. Only peer-reviewed articles were selected, given the available functionality of the chosen databases.

3. The search period ranged from 2002 to 2016. The rationale behind the start date was based on the pioneering work of two-sided platforms by Jean Rochet and Nobel Prize winner in Economics, Jean Tirole, in 2002 and 2003. In terms of management scholars, Anabelle Gawer and Michael Cusumano also published their seminal book Platform Leadership How Intel, Microsoft, and Cisco Drive Industry Innovation in 2002.
4. Only business-related subject areas were chosen (e.g. business economics, operations research, management science or information science), thus preventing the search results from convoluting, since both platform and service have a wide range of usage.

As a result of the limited functionalities and large variability of data from Google Scholar, it was excluded from the search. The following table shows the refined search results. The "filtered" line indicates the number of findings in each database after applying the inclusion and exclusion criteria (see Table 2).

Title and Abstract Screening

The abstract reviewing process further eliminated articles that were irrelevant to this literature review by focusing on two criteria. First, did the paper have a setting in the service-related context? Second, was the paper concerned with the two characteristics of the platform? After carefully reading the 1088 abstracts, 162 articles were selected for full paper review. Some of the abstracts required screening to clarify the subject area. The purpose was to clarify the ambiguous terms used in the abstracts. Finally, the remaining 133 papers formed the basis of this review. Figure 2 shows the selection process of key articles for this study.

Table 2 Core area "Service + Platform" search result

Criteria/databases	Science Direct	Web of Science
Not applied	3,499	2,736
Applied	331	1,088

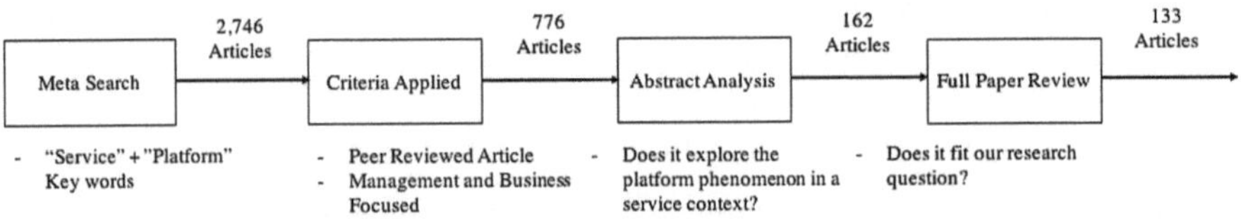

Fig. 2 Screening process

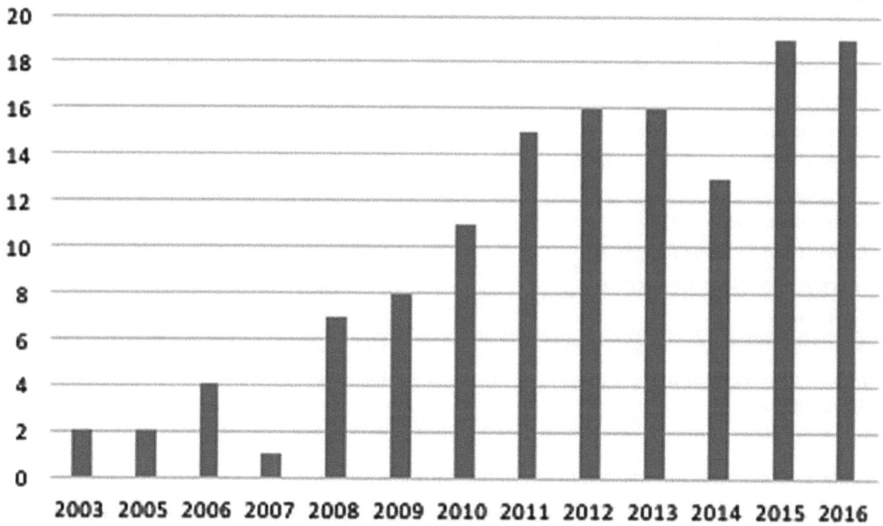

Fig. 3 Distribution of papers published annually

Descriptive Data

The 133 selected articles from the systematic literature review are analysed and presented in this section. The discussions and findings of this study are presented in the following section.

Research Distribution

In the early 2000s, the "platform" literature gained momentum, but it was not until 2008 that it gained significant attention (see Fig. 2). On closer inspection, two of the most cited papers in 2008 are "How companies become platform leaders", published in the MIT *Sloan Management Review,* and "How to sell service more profitably", which was published in the *Harvard Business Review*. Bridging the two phenomena may have become more relevant since then (Fig. 3).

The "service platform" topic attracts a wide array of interests from across disciplines. The literature is dispersed over 74 journals, with most of the publications being in the field of technology and information science. This was expected given that the root of the platform theory was inspired by earlier engineering and operation management concepts such as "modularity". Recently, however, management journals such as *Management Science* have started to publish on this topic. Table 3 shows the most popular journals that have published articles.

Table 3 Journals with more than two publications

Journal	No.	Journal	No.
Information & Management	5	*Telematics and Informatics*	3
Service Industries Journal	4	*Service Business*	2
Harvard Business Review	4	*MIS Quarterly*	2
Journal of Information Technology	4	*Journal of Electronic Commerce Research*	2
Telecommunications Policy	4	*Information Economics and Policy*	2
Information Systems Research	3	*Management Science*	2
Journal of Service Management	3	*Information Systems Journal*	2
Technovation	3	*Journal of Business Research*	2
Industrial Marketing Management	3	*MIT Sloan Management Review*	2
Information Systems and e-Business Management	3	*Marketing Science*	2

Table 4 Methodologies applied by the empirical studies

Methodology	No. of articles	Methodology	No. of articles
Case studies (multiple cases)	30	Survey	32
Case study (single case)	11	Secondary data analysis	13
Experiment	5		
Total			91

Given the infancy stage of "service platform" research, a considerable proportion of the work is conceptual. Approximately 68% (91 papers) of the research is made up of empirical studies with explicit data-gathering methodologies. A considerable portion of the research still comprises conceptual papers (Table 4).

The industries studied are consistent with the journal publications. In total, 55 studies were predominantly conducted within the IT or Internet-related industries, and most cases are set within the context of the social network (15 articles). The subjects include social media advertising, content services and e-word of mouth. E-business (11 articles) research on B2C and C2C commercial services follows closely after. The Internet is considered to be one of the key enablers of platform-based business (Table 5).

The topics on the service platform are also diverse. Appendix I shows a sample of the current research papers and a list of excerpts of some of the systematic literature review findings.

Findings and Trends

This section covers the findings from the literature in three areas. First, the theoretical foundation of the service platform is discussed, followed by the current research trends of the core literature. Finally, a few challenges, which are also potential research directions, are discussed.

Table 5 Top industries researched and subareas of IT industry

Industries researched	No. of articles
IT Internet	55
Telecom	14
Not specific	11
Service industry	10
Manufacturing	3
Retail	2
Subareas within IT industry	No. of articles
Social network	15
E-business	11
Crowdfunding	3
P2P services	3
Service-oriented architecture	3
Internet of things	2

Definition

Even though the service platform is a popular topic, as shown in this literature review, the definition of a service platform, or even "platform", is still being debated. This review has identified four main types of definition of platform: generic, architectural, economic and comprehensive, as shown in Fig. 4.

A significant portion of the papers included in this review have interpreted the term "platform" loosely. Some research uses the generic meaning of platform, indicating any online system as a "platform" (e.g. Cao et al. 2013). These papers do not discuss the modular architectural or economic features of the platforms. The focus of the papers typically evolves around aspects of services in the platform context. For instance, the studies of e-commerce platforms (e.g. Lehdonvirta 2009; Blasco-Arcas et al. 2014) go in depth to discuss user behaviours in the virtual marketplace. They emphasize cognitive drivers such as trust and service experience rather than network effects. However, these papers provide valuable insights for service platform researchers, as they offer alternative constructs to determine the performance of the service platform.

The platform-centric research accentuates two characteristics of service platforms. The first stream of literature is identified by the review as the "architectural aspect" of the platform research. In this context, a platform is defined as the common basis for product and service development (e.g. Gawer and Cusumano 2008). Gawer and Cusumano (2002) introduced the case of Intel's x86 chipset as a platform. External partners would join Intel's platform with their respective products, such as the video card by Nvidia, the hard drive by Western Data and the motherboard by ASUS, to provide the PC solution for the end customers. The profit from

PC customers is shared among these partners. This concept is derived from the modularity literature (Woodard and Baldwin, in Gawer 2009). Similar applications are also found in the service modularity literature (Pekkarinen and Ulkuniemi 2008).

The second stream of platform-centric literature is identified as the "economic aspect" of service platform research, which is mostly concerned with the network effect of platforms (e.g. Eisenmann et al. 2006). In other words, the more people engage in a platform, the more benefits are received by participants. Katz and Shapiro (1985) introduced the concept of network effect in their "network economics" work. The case of the telephone illustrates the value of the network. A single telephone does not generate any value for its user, since there is no one to call. However, the value of the telephone increases exponentially for every new phone introduced to the network.

Finally, a body of literature acknowledges both the architectural and economic aspects of the platform. Several authors have proposed that the theoretical foundation of the service platform requires more consolidation from the two aspects (e.g. Gawer 2014; Baldwin and Woodard 2009; Evans and Schmalensee 2007). This view has been adopted by an increasing number of authors, as shown in Fig. 4.

In terms of overall distribution, the architectural aspect is predominant. This is partially due to the literature on information technologies, where the emphasis of the research is on platform construction. However, there are a consistent number of publications that acknowledge a unified understanding of platforms (see Fig. 3). It is expected that more research will adopt a similar definition in the future, given the continued popularity of the research topic.

Specific definitions of the service platform are also emerging. In the area of service research, the service dominant logic (SDL) proposed by Vargo and Lusch (2008) has been widely cited. The current literature on service platforms has not extensively applied SDL in the context of the platform. Nevertheless, Lusch and Nambisan (2015) proposed the comprehensive application of SDL in the service platform context. The research landscape may be influenced in the future.

Discussion of Theoretical Foundation

Based on the platform-centric definitions, their theoretical foundation and directions for later research are discussed in this section. The architectural perspective of service platform research is partially inspired by the modularity research. In a modular system, each module fulfils a function and communicates with the others through standardized interfaces (Ulrich 1995). Contrary to the "integral" design, the components and functions have clear one-to-one relationship. Therefore, each part remains relatively independent from the other components. The "loose coupling" concept implies that the improved clarity and transparency of subsystems leads to many advantageous adjustments to complex systems (Campagnolo and Camuffo 2010).

Year Published	Generic	Architectural	Economic	Comprehensive	Total
2003		1	1		2
2005			2		2
2006	1	1		2	4
2007				1	1
2008		3	1	3	7
2009	1	1	2	4	8
2010	1	3		7	11
2011	3	8	2	2	15
2012	4	3	5	4	16
2013	3	4	1	8	16
2014	2	4	1	6	13
2015	5	5	4	5	19
2016	7	4	2	6	19
Grand Total	**27**	**37**	**21**	**48**	**133**

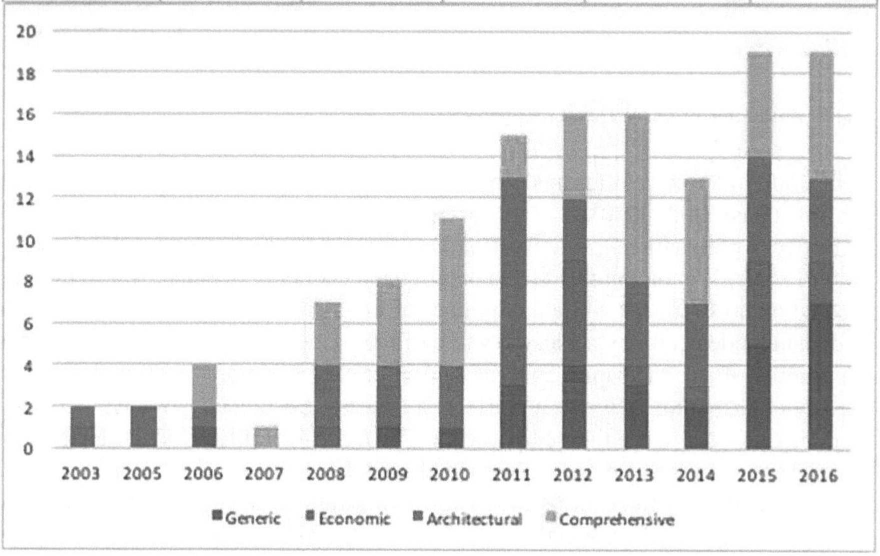

Fig. 4 Platform theoretical basis distribution annually

The most relevant architectural features can be summarized in the following three categories. First, duc to thc relative independence between each module, the engineers working on each module would enjoy a higher degree of freedom to allocate resources for new developments (Lau et al. 2010). Second, by sharing a common core platform, where the interface with the customers remain the same, the back-office operations can be modularized (Tuunanen and Cassab 2011). Therefore, aligning the strategic objective with existing resources would determine the most appropriate modules. Finally, by recombining the service modules, a higher degree

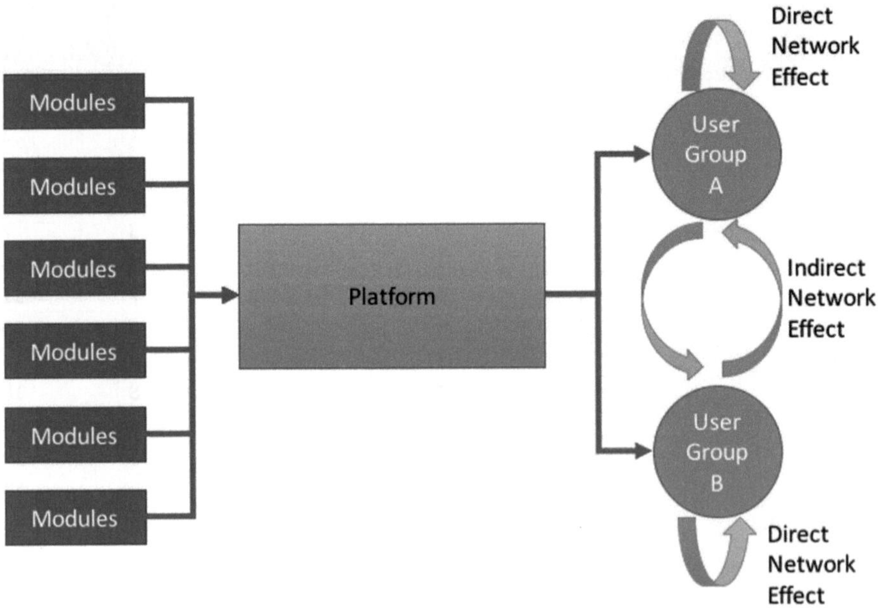

Fig. 5 Illustration of service platform features

of flexibility of service offerings can be achieved, which makes mass customization possible (Bask et al. 2010).

Network externalities or network effects can be referred as demand-side economy of scale. It is in contrast of the supply-side economy of scale, where the unit production cost reduces while the number of units produced increases. In the case of demand-side economy scale, the value of the product or service is contingent to the number of users (Shapiro and Varian 1998). Network effect is often deemed as the key contributing factor to a platform's success. Specifically two types of network effects exist in multi-sided platform. The right side of Fig. 5 illustrates a simplified version of the effects in a two-sided model.

Direct network effect refers to the effect that the number of the same type of users positively correlate with the total value of the product or service offered by the platform. In the case of social networking platforms such as Facebook, the more friends are signed up to the platform, the more valuable it becomes. Indirect network effect refers to the value creation among two or more groups of users. Indirect network effect can be both positive and negative. Positive indirect network effect can be exemplified by E-commerce platforms like eBay. The number of buyers would increase the value of the platform for sellers, who can benefit from a larger consumer base. Buyers on the other hand can benefit from more sellers with more variety of products. Negative indirect network effect occurs when the complementarity of the two sides of the platform misalign. For example, the matchmaking website between men and women can seize to be valuable when one side of the subscribers overwhelm the other side. If the number of men far supersedes women by a large margin, women tend to be overwhelmed by the number of males seeking to connect. The women users can be disturbed by the information overload. Men on the other hand may find lack of success in trying to connect with women discouraging, which consequently render the platform worthless. These negative effects can be offset by utilizing strategies such as pricing one side and subsidizing the other (e.g. Bhargava et al. 2013).

Based on the above characteristics, a wide spectrum of researches have been conducted in the service context. These researches are summarized in the section below. A list of short descriptions of these studies in the appendix can also serve as guide to the research area.

Three Categories of Research

"Service" and "platform" cover a wide spectrum of topics. Based on the papers' perspective of service platforms, this review divides the literature into three broad categories: (1) service architecture/modularity, (2) the platform's impact on services and (3) service platform strategy. The taxonomy is shown in Fig. 6.

The first category consists of research that applies "platform" thinking to the field of services. This category is referred to as "service architecture/modularity". Much conceptual work has been conducted, and the amount of research has been increasing. However, empirical research is still limited. Prior to 2008, only one paper was published on service modularity (Bask et al. 2010; Pekkarinen and Ulkuniemi 2008). The studies are based on the service industry. As a result of the fact that services in the traditional sense tend to adjust their offerings according to customer requests, a satisfactory degree of service modularity has not been observed (Bask et al. 2010). Pekkarinen and Ulkuniemi attempted to construct a model for service modularity. However, their research is based on the single case of a logistics service provider. The validity of their proposed model therefore requires further examination. Tuunanen and Cassab (2011) conducted a controlled experiment to determine the service process module reusability against the complexity of service, which sheds light on the research direction. However, the causes of low architectural leverage of platform capabilities in the service industry are still unclear, even though platform and modular design concepts in the service industry have not generated significant momentum.

The second category of literature focuses on how the platform has changed the way companies run their business. We have named this category the "platform's impact". This topic covers a wide spectrum of activities from innovation and operations to marketing and industrial architecture. Two subcategories have been identified. The first subcategory, "market disruption", consists of papers discussing how the introduction of the platform in the service industry has changed how service professionals conduct business. Seamans and Zhu (2014) discussed how Craigslist has influenced the newspaper industry. The second subcategory is called "service platform market condition". Craigslist has shifted the revenue model of the

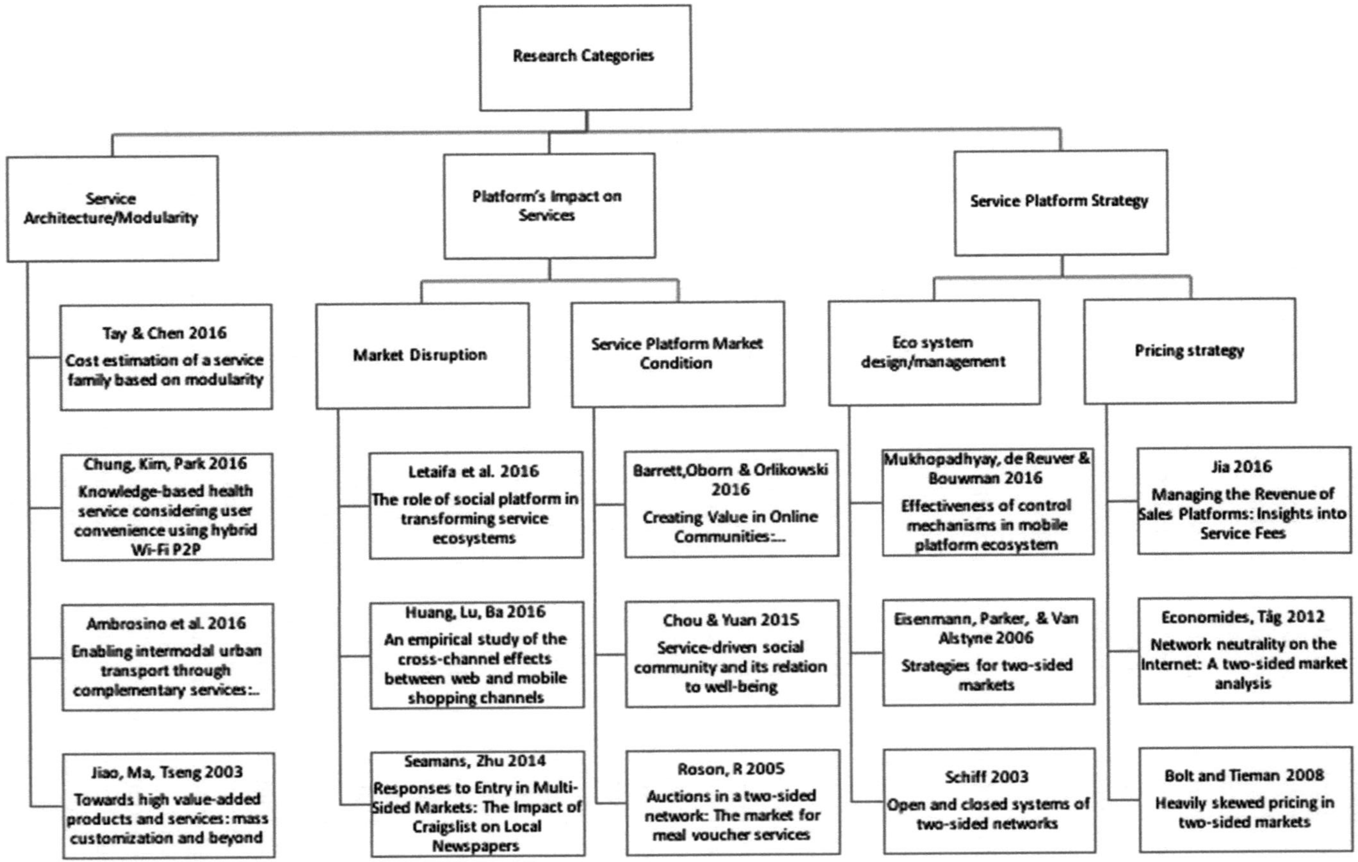

Fig. 6 Categories of research and sample publications

newspapers away from relying on targeted listing advertisings to subscription fees. These studies focus on how the service platform functions, without deliberately discussing the platform's architecture or network effects. Weiss and Gangadharan (2010) suggest that the innovation patterns of app service providers in the platform context differ from the traditional ones. Rather than expanding the breadth of services, they tend to focus on a particular type of app and increase its "depth" or volume within a narrow scope. Reisiger et al. (2009) take radio stations to be a two-sided platform and analyse the relationship between advertisers and radio service consumers. This stream of literature provides a rich understanding of the platform business. However, because of the wide spectrum of topics in this research stream, consensus among scholars on methodologies, concepts or research directions is rare. Nevertheless, the explorative studies are valuable in terms of determining important research questions for future research.

The final category shows the most prominent research directions for platform literature, evolving around what makes a company a platform leader and how a company can maintain its leadership position (Gawer and Cusumano 2008). The metric of leadership could be interpreted in many ways, for example, monetary, customer value and market share. Several empirical researchers have suggested that much of the information, such as financial data or customer value, is very hard to obtain or objectively determine; therefore, the most reasonable metric for the current platform research tends to focus on the number of users (Evans and Schmalensee 2010; Lin et al. 2012). This measure is also in accordance with the principles of network effect that the growth in the number of users increases network externality.

To achieve platform leadership, researchers have focused on the two characteristics of platforms, namely, how to leverage the technology core of the platform, known as "coring", and how to leverage the platform's network externality, known as "tipping strategy" (Lee et al. 2010; etc.). Using strategies from the technology side of the spectrum, a platform leader creates a high level of entry barrier for potential challengers. For example, Intel invests heavily in its microchip technology, which makes potential entry into the microprocessor platform more difficult. To leverage the network effects, platform owners usually create incentives to encourage network participants. This could be in the form of benefits for customers or providers. YouTube subsidizes its content providers by sharing advertising revenue generated by visitor traffic. Recent research has shown some promising strategies to maintain platform users through governance. Eaton et al. (2015) analysed the iOS platform and app offerings by encouraging certain types of offering and limiting others. Apple achieved higher customer satisfaction and therefore retention rate.

Challenges and Opportunities

The first challenge arrives from the advancements of Internet technology. Compared with previous studies of platforms with distinct psychical technologies, such as video cassette players and game consoles, in the setting of digital service platforms such as Uber and AirBnB, very few sunk costs, such as equipment purchase prices, are imposed on customers. From a transactional cost perspective, many information goods and services have virtually zero marginal costs (Gawer 2014). On the other hand, the "core" technologies provided by these newly emerged platforms are not particularly hard to create, given the ease of programming the modern Web and mobile technologies (Kim et al. 2012).

The second challenge is the adoption issue, which is characterized as a chicken-and-egg problem. One commonly agreed notion of platform network externality is that the increase in the variety and quality of product and service offerings tends to attract customers (Boudreau 2012; Hsieh and Hsieh 2013). The network externalities are dependent on both sides of the market; without a large enough customer base, providers are unlikely to join and innovate, and without enough offerings available, customers will not materialize (Eisenmann and Hagiu 2007). The current literature suggests a solution to the issue through capabilities (e.g. Tan et al. 2015), pricing (e.g. Bolt and Tieman 2008; Hagiu 2009), strategic alliances (e.g. Caesy and Toyli 2012) or ecosystem value co-creation (e.g. Ceccagnoli et al. 2012). Little research has focused on appealing to the provider side of the market (Hsieh and Hsieh 2013).

The strategy literature on the platform has highlighted the subsidizing supply side as a method to sustain platform leadership. For example, Intel could convince motherboard makers to adopt their PCI standard by committing its own microprocessor production volume (Gawer and Cusumano 2007). However, a substantial study of the factors that influence providers' adoption and innovation decisions is currently unavailable. As mentioned earlier, platform customers have very low sunk costs, which also reduces the switch cost and undermines the lock-in effect. The same applies to the provider side of the platform. Providers have also shown strong incentives to switch between platforms given the right circumstances (Lin et al. 2012).

Finally, the current research agenda of the platform with respect to adoption is generally limited to the economic and technological rationale of the platform strategy (Thomas et al. 2014). Recent research has pointed to areas of cognitive biases, such as the "bandwagon effect", which have been put into the research agenda (Xu et al. 2012). However, the results of this research have not been tested on a wider scale. On the one hand, some researchers have taken into consideration the intangible aspects of platform strategies. This type of research is still at an innate stage, and a systematic understanding of the platform adoption process is missing. On the other hand, marketing researchers have studied customer behaviour from a non-economic perspective. Phenomena such as word of mouth (Shin et al. 2014) and viral marketing (Palka et al. 2009), even B2C communication via sponsored messages (Magnini 2011), can have a significant impact on platform users' behaviour. There is a large research gap in terms of the study of complementors or service innovation contributors.

Many platform providers understand that the importance of platform success in constructing a meaningful business model relies heavily on the sheer number of participants. Therefore, in many cases of Internet-based platforms, the content is offered free of charge. Scholars consider pricing and access limitations to be potentially useful tools in terms of quality control (Economides and Hermalin 2015). Furthermore, platform leaders such as Apple tend to be able to manage the quality of the content of their platforms through the governance of boundary resources (Eaton et al. 2015). However, further studies on the quality aspects of the platform are not widely covered. Therefore, it would be particularly meaningful to understand what drives providers in a platform to innovate quality services.

Discussion and Conclusions

This systematic literature review was carried out on the service platform. This chapter provides a holistic overview of the current situation regarding this subject. The review shows that research on the service platform increased rapidly after 2008. A wide spectrum of research from different industries, methodologies and scientific disciplines has been covered. Despite the increasing interests in the service platform area in recent years, there are still many areas to be explored.

This chapter has identified the need for a better and more comprehensive theoretical foundation for the literature on the service platform. A converging view of platform has been observed among management scholars. However, the implication of services in the platform context has not been clearly identified. Comparative studies between service and technological platforms may shed light to further strengthen our understanding of the core concepts.

The service architecture research agenda needs to be further perused with more empirical data support. Many technical architectures have been proposed in the service contexts. However, a critical evaluation of such models based on longitudinal studies of multiple cases is still rare. A "dominant logic" of service platform architecture has not yet been observed in this stream of research.

Finally, despite a great number of directions that strategic management scholars have embarked on studying service platforms, some fundamental questions are still worth perusing. Among those, the "chicken and egg" issue of platform adoption is still central to the success of launching a platform. Current theories on adoption are mostly descriptive of the key stages of platform user growth, which tend to offer little predictive power. More fundamental causes of customer adoption need to be examined. Another aspect concerning customer loyalty of service platform may need further exploration. Platform-based services often tend to become "commoditized", where customers show little loyalty in switching between the service providers. How a service platform can compete in terms of value proposition beyond matchmaking is an interesting and critical question.

In summary, we believe it is both timely and important to conduct this literature review on service platforms. This review identifies the current research streams and updates the research agenda. This provides exciting opportunities for management scholars to advance our understanding of service platforms. It is also valuable for readers in industry to identify their business' potential benefits and challenges from service platforms. An increasing number of firms are seeking to engage in service platforms. This comprehensive review of the cutting-edge researches and case studies can be used by organizations as a key reference when approaching service platforms.

Appendix I: Snippets of Publications

Authors	Year	Summary
Barrett, Michael; Oborn, Eivor; Orlikowski, Wanda	2016	The authors conducted a longitudinal field study of a health-care social platform. They identified a complex network in which the online community value is orchestrated
Tay, Choon Khai; Chen, Song Lin	2016	This paper presents a cost estimation model for the service family based on modularity
Lusch, Robert F.; Nambisan, Satish	2015	Service platforms, which enhance the efficiency and effectiveness of the service exchange by liquefying resources and increasing resource density
Hofman, Erwin; Meijerink, Jeroen	2015	This study finds that the service value is highest when the service provision is matched with the commonality potential of the services. The results indicate that using the wrong delivery channel decreases the service value, which could eventually decrease the service value for an organization's external customers
Eaton, Ben; Elaluf-Calderwood, Silvia; Sorensen, Carsten	2015	The tuning of 30 boundary resources can influence the innovation dynamics of the iOS platform
Seamans, Robert; Zhu, Feng	2014	Relative to newspapers without classified ad managers, the effect of Craigslist's entry on newspapers with classified ad managers has led to a decrease of 20.7% in classified ad rates, an increase of 3.3% in subscription prices, a decrease of 4.4% in circulation, an increase of 16.5% in differentiation and a decrease of 3.1% in display ad rates. Craigslist's entry has decreased the attractiveness of the newspaper to classified advertisers, which now have an alternative channel to reach newspaper subscribers. As a result, the affected newspaper decreases the classified ad rate. The newspaper now has a lower incentive to subsidize the subscriber side because each eyeball no longer generates the same amount of ad revenue as before, a finding that is consistent with the existing theory (e.g. Godes et al. 2009; Hagiu 2009)

(continued)

Authors	Year	Summary
Gawer, Annabelle; Cusumano, Michael A.	2014	This paper defines the distinction between internal and external platforms and emphasizes the importance of network externalities in these platforms. The Intel case is used to illustrate an ecosystem platform leader. A comparative study among IBM, Intel and Microsoft discusses their evolution towards platform leaders. A study centring on Google and Nokia in the mobile phone industry and a comparison between Microsoft and Apple in the software industry were used to further strengthen the evolution trajectories of the platform leaders and losers
Pon, Bryan; Seppala, Timo; Kenney, Martin	2014	This paper describes the transition in the mobile industry, where the device as a key asset to ensure a healthy ecosystem is no longer valid, and companies such as Google, with its Android platform, are proposing a new paradigm. Previous strategies to compete with operation systems are no longer relevant. This paper also analyses the gatekeeper roles of three such ecosystems in terms of the service-creation environment, identity management, service provisions and billing
Chen, Dongyu; Lai, Fujun; Lin, Zhangxi	2014	This paper examines lender's behaviours in p2p lending platforms. The authors proposed a model that user's trust in intermediary and trust in borrower would determine a platform user likelihood of lending. These two types of trust are based on 5 specific factors, which are familiarity of the platform, service quality, security protection, social capital and information quality.
Battistella, Cinzia; Nonino, Fabio	2013	This research establishes the relationships among the intrinsic/ extrinsic motivations and the likelihood of using open innovation Web-based platforms
Hagiu, Andrei; Wright, Julian	2013	This paper identifies challenges for platform wannabes, such as sales efficiency, network size and competition
Bhargava, Hemant K.; Kim, Byung Cho; Sun, Daewon	2013	The authors propose a model to predict the optimal expansion strategies for start-ups and established firms to benefit from network externalities
Suarez, Fernando F.; Kirtley, Jacqueline	2012	Executive summary of four strategies: (1) Target an under-served segment of the overall customer base. (2) Leverage adjacent platforms to boost demand. (3) Differentiate their product to meet emerging needs. (4) Expand the universe of potential partners by simplifying the business model for partners
Madni, Azad M.	2012	Identifies trends of platform-based engineering, suggesting a more resilient and flexible adaptable PBE framework to avoid platform "lock-in", especially in the engineering aspect in the long term
Lu, June; Wang, Luzhuang; Hayes, Linda A.	2012	Optimism and insecurity influence the C2C platform's trust and functionality, which ultimately influence C2C satisfaction
Casey, Thomas R.; Toyli, Juuso	2012	The authors introduce a system-dynamics-based theoretical model to simulate the adoption of public and local wireless platforms. This paper highlights the importance of understanding complex feedback loops of the value network

(continued)

Authors	Year	Summary
Beltran, Fernando	2012	The authors use a two-sided platform concept to analyse the UFB market in New Zealand. The results show that a multi-sided market approach is beneficial to end customers, and network neutrality can be a catalyst for the deployment of broadband by combining access market and content market
Kim, Jieun; Lee, Sungjoo; Geum, Youngjung; Park, Yongtae	2012	The structure of digital content services using three building blocks: product, process and platform. Basically, two types of innovation pattern are established as divergence and convergence
Yoo, Youngjin; Boland, Richard J., Jr.; Lyytinen, Kalle; Majchrzak, Ann	2012	This paper summarizes (1) the importance of digital technology platforms, (2) the emergence of distributed innovations and (3) the prevalence of combinatorial innovation. Digital platforms are a means to promote distributed recombination innovations
Tuunanen, Tuure; Cassab, Harold	2011	Contingent to task complexity, modularized services positively influence customers' perceived value of services and their likelihood to engage in trials of service extensions
Kaplan, Andreas M.; Haenlein, Michael	2011	Identifies three justifications for micro-blogging to exist: virtual exhibitionism and voyeurism, pre-purchase marketing research and post-purchase customer relationship management
Luis Osorio, A.; Afsarmanesh, Hamideh; Camarinha-Matos, Luis M.	2011	Proposes a framework for integrating services
Jung, Jason J.	2011	Proposes a possible service-oriented architecture-enabled SC structure
Liang, Ting-Peng; Ho, Yi-Ting; Li, Yu-Wen; Turban, Efraim	2011	Service quality, relationship quality and website quality influence social commerce decisions
Shang, Shari S. C.; Li, Eldon Y.; Wu, Ya-Ling; Hou, Oliver C. L.	2011	Taxonomy of Web 2.0 service models introduced based on the knowledge-creation perspective
Zoric, Josip	2011	This paper introduces a techno-business modelling approach concerning the business model in a service platform design scenario. Using models from his work of 2010, he proposes scenario planning by mapping the services, service enablers, capabilities and resources to address the appropriate service platform design that is fit for purpose
Moon, Seung Ki; Shu, Jun; Simpson, Timothy W.; Kumara, Soundar R. T.	2011	This paper presents a module-based service model for mass customization. The model has a three-phase design: (1) service process identification, (2) service platform design and (3) platform strategy determination

(continued)

Authors	Year	Summary
Weiss, Michael; Gangadharan, G. R.	2010	(1) A small number of APIs provide the basis for the majority of mashups. (2) Mashup platforms were introduced in response to the increasing complexity of mashups, as mashups evolved from one-feature mashups (widgets). (3) The growth of the mashup ecosystem follows a pattern where keystone data providers or "powerful hubs" attract niche data providers as complementors, and the positions of keystones in the ecosystem are mutually reinforcing
Evans, David S.; Schmalensee, Richard	2010	(1) Platform businesses typically need to attain critical mass when they are launched in order just to survive. (2) With direct network effects, the basic problem is that the level of participation inthe platform affects the quality of the product it offers to participants; if the quality is too low, participation falls, which reduces the quality further, with participation declining towards zero
Luo, Xin; Li, Han; Zhang, Jie; Shim, J. P.	2010	Risk perception influences the adoption of Internet banking services
Beeflamme, Paul; Peitz, Martin	2010	This paper proposes a model that predicts the incentives for open platform sellers. For instance, in the two-sided single-homing environment, sellers would have more incentives to invest, whereas if the buyers were multi-homing, sellers would be less likely to invest
Lehdonvirta, Vili	2009	Hedonic and social attributes act as purchase drivers
Reisinger, Markus; Ressner, Ludwig; Schmidtke, Richard	2009	In addition to considering participation externality, *pecuniary* externality, such as the revenue stream, can also have an influence on the dynamics of a two-sided platform. The research takes radio stations as a two-sided platform and analyses the relationship between advertisers and radio service consumers
Tee, Richard; Gawer, Annabelle	2009	Industry structure can be a determining factor for the success or failure of mobile Internet services
Hagiu, Andrei; Yoffie, David B.	2009	MSP does not guarantee a participant's success; a company should align its product and service offerings with the type of platform activities in which it should engage
Reinartz, Werner; Ulaga, Wolfgang	2008	A flexible service platform can help companies to sell services more profitably and potentially achieve higher customer satisfaction
Lai, Linda S. L.; Turban, Efraim	2008	There is a positive correlation between network value and user content. Services are defined in general terms, such as self-serving services and collaboration services (Google talk)
Pekkarinen, Saara; Ulkuniemi, Pauliina	2008	Service modularity's success depends on the manager's knowledge in terms of choosing the correct processes and coordinating such modules to the organizations
Bolt, Wilko; Tieman, Alexander F.	2008	The most elastic side of the market is used to generate maximum demand by providing it with platform services at the lowest possible price. Full participation of the high-elasticity, low-price side of the market attracts the other side. As this side is less price-elastic, the platform is able to extract high prices
Gawer, Annabelle; Cusumano, Michael A.	2008	Through a series of cases from several industries, the authors offer strategic guidance for companies to achieve platform leadership

(continued)

References

Albuquerque, P., Pavlidis, P., Chatow, U., Chen, K.-Y., & Jamal, Z. (2012). Evaluating promotional activities in an online two-sided market of user-generated content. *Marketing Science, 31*(3), 406–432.

Ambrosino, G., Nelson, J. D., Boero, M., & Pettinelli, I. (2016). Enabling intermodal urban transport through complementary services: From flexible mobility services to the shared use mobility agency workshop 4. Developing inter-modal transport systems. *Research in Transportation Economics, 59*(0), 179–184.

Aulkemeier, F., Paramartha, M. A., Iacob, M.-E., & van Hillegersberg, J. (2016). A pluggable service platform architecture for e-commerce. *Information Systems and e-Business Management, 14*(3), 469–489.

Avgerou, C., & Li, B. (2013). Relational and institutional embeddedness of web-enabled entrepreneurial networks: Case studies of netrepreneurs in China. *Information Systems Journal, 23*(4), 329–350.

Balaji, M. S., Khong, K. W., & Chong, A. Y. L. (2016). Determinants of negative word-of-mouth communication using social networking sites. *Information Management, 53*(4), 528–540.

Baldwin, C. Y., & Jason Woodard, C. (2008). The architecture of platforms: A unified view. *Working Papers – Harvard Business School Division of Research*, 1–31.

Banoun, A., Dufour, L., & Andiappan, M. (2016). Evolution of a service ecosystem: Longitudinal evidence from multiple shared services centers based on the economies of worth framework. *Journal of Business Research, 69*(8), 2990–2998.

Barrett, M., Oborn, E., & Orlikowski, W. (2016). Creating value in online communities: The sociomaterial configuring of strategy, platform, and stakeholder engagement. *Information Systems Research, 27*(4), 704–723.

Bask, A., Tinnilä, M., & Rajahonka, M. (2010). Matching service strategies, business models and modular business processes. *Business Process Management Journal, 16*(1), 153–180.

Blasco-Arcas, L., Hernandez-Ortega, B., & Jimenez-Martinez, J. (2014). The online purchase as a context for co-creating experiences. Drivers of and consequences for customer behavior. *Internet Research, 24*(3), 393–412.

Battistella, C., & Nonino, F. (2013). Exploring the impact of motivations on the attraction of innovation roles in open innovation web-based platforms. *Production Planning & Control, 24*(42065), 226–245.

Beeflamme, P., & Peitz, M. (2010). Platform competition and seller investment incentives. *European Economic Review, 54*(8), 1059–1076.

Beltran, F. (2012). Using the economics of platforms to understand the broadband-based market formation in the New Zealand Ultra-Fast Broadband Network. *Telecommunications Policy, 36*(9), 724–735.

Ben Letaifa, S., Edvardsson, B., & Tronvoll, B. (2016). The role of social platforms in transforming service ecosystems. *Journal of Business Research, 69*(5), 1933–1938.

Bhargava, H. K., Kim, B. C., & Sun, D. (2013). Commercialization of platform technologies: Launch timing and versioning strategy. *Production and Operations Management, 22*(6), 1374–1388.

Blasco-Arcas, Lorena, Blanca Hernandez-Ortega, and Julio Jimenez-Martinez. 2014. "The Online Purchase as a Context for Co-Creating Experiences. Drivers of and Consequences for Customer Behavior." Internet Research 24(3): 393–412.

Bolt, W., & Tieman, A. F. (2008). Heavily skewed pricing in two-sided markets. *International Journal of Industrial Organization, 26*(5), 1250–1255.

Boudreau, K. J. (2012). Let a thousand flowers bloom? An early look at large numbers of software app developers and patterns of innovation. *Organization Science, 23*(5), 1409–1427.

Breeding, M. (2012). Current and future trends in information technologies for information units. *Profesional De La Informacion, 21*(1), 9–15.

Campagnolo, D., & Camuffo, A. (2010). The concept of modularity in management studies: A literature review. *International Journal of Management Reviews, 12*, 259–283.
Cao, H., Jiang, J., Oh, L.-B., Li, H., Liao, X., & Chen, Z. (2013). A Maslow's hierarchy of needs analysis of social networking services continuance. *Journal of Service Management, 24*(2), 170–190.
Casalo, L. V., Flavian, C., & Guinaliu, M. (2010). Relationship quality, community promotion and brand loyalty in virtual communities: Evidence from free software communities. *International Journal of Information Management, 30*(4), 357–367.
Casey, T. R., & Toyli, J. (2012). Dynamics of two-sided platform success and failure: An analysis of public wireless local area access. *Technovation, 32*(12), 703–716.
Ceccagnoli, M., Forman, C., Huang, P., & Wu, D. J. (2012). Cocreation of value in a platform ecosystem: The case of enterprise software. *MIS Quarterly, 36*(1), 263–290.
Cerf, V. G. (2008). Innovation and the Internet. *Research-Technology Management, 51*(1), 30–33.
Chen, W.-T., & Hu, C.-N. (2008). Entering the mobile service market via mobile platforms: Qualcomm's BREW platform and Nokia's Preminet platform. *Telecommunications Policy, 32*(6), 399–411.
Chen, K.-P., & Huang, Y.-C. (2012). A search-matching model of the buyer-seller platforms. *CESifo Economic Studies, 58*(4), 626–649.
Chen, T.-Y., Tsaih, D., & Chen, Y.-M. (2010). A knowledge-commercialised business model for collaborative innovation environments. *International Journal of Computer Integrated Manufacturing, 23*(6), 543–564.
Chen, D., Lai, F., & Lin, Z. (2014a). A trust model for online peer-to-peer lending: A lender's perspective. *Information Technology & Management, 15*(4), 239–254.
Chen, J., Chen, J. (Elaine), Goh, K. Y., Xu, Y. (Calvin), & Tan, B. C. Y. (2014b). When do sellers bifurcate from Electronic Multisided Platforms? The effects of customer demand, competitive intensity, and service differentiation. *Information & Management, 51*(8), 972–983.
Chiang, I. R., & Jhang-Li, J.-H. (2014). Delivery consolidation and service competition among internet service providers. *Journal of Management Information Systems, 31*(3), 254–286.
Choi, J. P., Jeon, D.-S., & Kim, B.-C. (2015). Net neutrality, business models, and internet interconnection. *American Economic Journal: Microeconomics, 7*(3), 104–141.
Chou, C. Y., & Yuan, S.-T. (2015). Service-driven social community and its relation to well-being. *Service Industries Journal, 35*(42954), 368–387.
Chung, K., Kim, J.-C., & Park, R. C. (2016). Knowledge-based health service considering user convenience using hybrid Wi-Fi P2P. *Information Technology & Management, 17*(1), 67–80.
Dario Franco, R., Ortiz Bas, A., & Lario Esteban, F. (2009). Modeling extended manufacturing processes with service-oriented entities. *Service Business, 3*(1), 31–50.
de Reuver, M., Verschuur, E., Nikayin, F., Cerpa, N., & Bouwman, H. (2015). Collective action for mobile payment platforms: A case study on collaboration issues between banks and telecom operators. *Electronic Commerce Research and Applications, 14*(5), 331–344.
Eaton, B., Elaluf-Calderwood, S., & Sorensen, C. (2015). Distributed tuning of boundary resources: The case of Apple's iOS service system. *MIS Quarterly, 39*(1), 217–243.
Economides, N., & Hermalin, B. E. (2015). The strategic use of download limits by a monopoly platform. *RAND Journal of Economics, 46*(2), 297–327.
Economides, N., & Tag, J. (2012). Network neutrality on the Internet: A two-sided market analysis. *Information Economics and Policy, 24*(2), 91–104.
Edvardsson, B., & Enquist, B. (2011). The service excellence and innovation model: Lessons from IKEA and other service frontiers. *Total Quality Management & Business Excellence, 22*(5), 535–551.
Eisenmann, T., Parker, G., & Van Alstyne, M. W. (2006). Strategies for two-sided markets. *Harvard Business Review, 84*(10), 92.
Eisenmann, T. R., & Hagiu, A. (2007). *Staging two-sided platforms.*
Eloranta, V., & Turunen, T. (2016). Platforms in service-driven manufacturing: Leveraging complexity by connecting, sharing, and integrating. *Industrial Marketing Management, 55*(0), 178–186.

Evans, D. S. (2013). Attention rivalry among online platforms. *Journal of Competition Law & Economics, 9*(2), 313–357.

Evans, D. S., & Schalensee, R. (2007). The industrial organization of markets with two-sided platforms. *Competition Policy International, 3*(1), 150–179.

Evans, D. S., & Schmalensee, R. (2010). Failure to launch: Critical mass in platform businesses. *Review of Network Economics, 9*(4), 0–0.

Evans, D. S., & Schmalensee, R. (2016). *Matchmakers: The new economics of multisided platforms*. Boston: Harvard Business Review Press.

Evans, S., Partidario, P. J., & Lambert, J. (2007). Industrialization as a key element of sustainable product-service solutions. *International Journal of Production Research, 45*(18–19), 4225–4246.

Ezema, I. J., Ezeah, C. S., & Ishiwu, B. N. (2015). Social networking services: A new platform for participation in government programmes and policies among Nigerian youths. *Libres-Library and Information Science Research Electronic Journal, 25*(1), 33–49.

Gawer, A. (2009). Platform dynamics and strategies: From products to services. *Platforms, markets and innovation, 45*, 57.

Gawer, A. (2011). *Platforms, markets and innovation*. Cheltenham: Edward Elgar Publishing.

Gawer, A. (2014). Bridging differing perspectives on technological platforms: Toward an integrative framework. *Research Policy, 43*(7), 1239–1249.

Gawer, A., & Cusumano, M. A. (2002). *Platform leadership: How Intel, Microsoft, and Cisco drive industry innovation* (Vol. 5, pp. 29–30). Boston: Harvard Business School Press.

Gawer, A., & Cusumano, M. (2007, June). A strategy toolkit for platform leader wannabes. In *DRUID summer conference* (pp. 1–33).

Gawer, A., & Cusumano, M. A. (2008). How companies become platform leaders. *MIT Sloan Management Review, 49*(2), 28.

Gawer, A., & Cusumano, M. A. (2014). Industry platforms and ecosystem innovation. *Journal of Product Innovation Management, 31*(3), 417–433.

Gedda, D., Nilsson, B., Sathen, Z., & Soilen, K. S. (2016). Crowdfunding: Finding the optimal platform for funders and entrepreneurs. *Technology Innovation Management Review, 0*(0), 31–40.

Gnyawali, D. R., Fan, W., & Penner, J. (2010). Competitive actions and dynamics in the digital age: An empirical investigation of social networking firms. *Information Systems Research, 21*(3), 594–613.

Godes, D., & Mayzlin, D. (2009). Firm-created word-of-mouth communication: Evidence from a field test. *Marketing Science, 28*(4), 721–739.

Goncalves, V., & Ballon, P. (2011). Adding value to the network: Mobile operators' experiments with Software-as-a-Service and Platform-as-a-Service models. *Telematics and Informatics, 28*(1), 12–21.

Grotnes, E. (2009). Standardization as open innovation: Two cases from the mobile industry. *Information Technology & People, 22*(4), 367–381.

Hagiu, A., & Wright, J. (2013). Do you really want to be an eBay? *Harvard Business Review, 91*(3), 102.

Hagiu, A., & Yoffie, D. B. (2009). What's your Google strategy? *Harvard Business Review, 87*(4), 74.

Hagiu, A., & Wright, J. (2015). Multi-sided platforms. *International Journal of Industrial Organization, 43*, 162–174.

Hagiu, A. (2009). Two-sided platforms: Product variety and pricing structures. *Journal of Economics and Management Strategy, 18*(4), 1011–1043.

Hofman, E., & Meijerink, J. (2015). Platform thinking for services: The case of human resources. *Service Industries Journal, 35*(3), 115–132.

Hsieh, J.-K., & Hsieh, Y.-C. (2013). Appealing to Internet-based freelance developers in smartphone application marketplaces. *International Journal of Information Management, 33*(2), 308–317.

Huang, P., Ceccagnoli, M., Forman, C., & Wu, D. J. (2013). Appropriability mechanisms and the platform partnership decision: Evidence from enterprise software. *Management Science, 59*(1), 102–121.

Huang, L., Lu, X., & Ba, S. (2016). An empirical study of the cross-channel effects between web and mobile shopping channels. *Information Management, 53*(2), 265–278.

Janssen, M., & Zuiderwijk, A. (2014). Infomediary business models for connecting open data providers and users. *Social Science Computer Review, 32*(5), 694–711.

Jarmolowicz, D. P., Bickel, W. K., Carter, A. E., Franck, C. T., & Mueller, E. T. (2012). Using crowdsourcing to examine relations between delay and probability discounting. *Behavioural Processes, 91*(3), 308–312.

Jia, J. (2016). Managing the revenue of sales platforms: Insights into service fees. *Production and Operations Management, 25*(6), 1026–1037.

Jiang, B., Jerath, K., & Srinivasan, K. (2011). Firm strategies in the "mid tail" of platform-based retailing. *Marketing Science, 30*(5), 757–775.

Jiao, J. X., Ma, Q. H., & Tseng, M. M. (2003). Towards high value-added products and services: Mass customization and beyond. *Technovation, 23*(10), 809–821.

Jung, J. J. (2011). Service chain-based business alliance formation in service-oriented architecture. *Expert Systems with Applications, 38*(3), 2206–2211.

Kaplan, A. M., & Haenlein, M. (2011). The early bird catches the news: Nine things you should know about micro-blogging. *Business Horizons, 54*(2), 105–113.

Karippacheril, T. G., Nikayin, F., de Reuver, M., & Bouwman, H. (2013). Serving the poor: Multisided mobile service platforms, openness, competition, collaboration and the struggle for leadership. *Telecommunications Policy, 37*(1), 24–34.

Katz, M. L., & Shapiro, C. (1985). Network externalities, competition, and compatibility. *The American Economic Review, 75*(3), 424–440.

Ke, D., Chen, A., & Su, C. (2016). Online trust-building mechanisms for existing brands: The moderating role of the e-business platform certification system. *Electronic Commerce Research, 16*(2), 189–216.

Kijima, K., Rintamki, T., & Mitronen, L. (2013). Value orchestration platform: Model and strategies. *Advances in the Human Side of Service Engineering, 0*(0), 433–442.

Kim, D. (2016). Value ecosystem models for social media services. *Technological Forecasting and Social Change, 107*(0), 13–27.

Kim, J., Lee, S., Geum, Y., & Park, Y. (2012). Patterns of innovation in digital content services: The case of App Store applications. *Innovation: Management Policy & Practice, 14*(4), 540–556.

Kuk, G., & Janssen, M. (2013). Assembling infrastructures and business models for service design and innovation. *Information Systems Journal, 23*(5), 445–469.

Lahiri, A., Dewan, R. M., & Freimer, M. (2010). The disruptive effect of open platforms on markets for wireless services. *Journal of Management Information Systems, 27*(3), 81–109.

Lai, L. S. L., & Turban, E. (2008). Groups formation and operations in the web 2.0 environment and social networks. *Group Decision and Negotiation, 17*(5), 387–402.

Lau, A. K. W., Tang, E., & Yam, R. C. M. (2010). Effects of supplier and customer integration on product innovation and performance: Empirical evidence in Hong Kong manufacturers. *Journal of Product Innovation Management, 27*(5), 761–777.

Lee, S. M., Kim, T., Noh, Y., & Lee, B. (2010). Success factors of platform leadership in web 2.0 service business. *Service Business, 4*(2), 89–103.

Lehdonvirta, V. (2009). Virtual item sales as a revenue model: Identifying attributes that drive purchase decisions. *Electronic Commerce Research, 9*(1-2), 97–113.

Li, Y.-M., & Lee, Y.-L. (2010). Pricing peer-produced services: Quality, capacity, and competition issues. *European Journal of Operational Research, 207*(3), 1658–1668.

Li, Z., Huang, G. Q., Fang, J., & Qu, T. (2014). Ontology-based dynamic alliance services (ODAS) in production service system. *International Journal of Computer Integrated Manufacturing, 27*(2), 148–164.

Liang, T.-P., Ho, Y.-T., Li, Y.-W., & Turban, E. (2011). What drives social commerce: The role of social support and relationship quality. *International Journal of Electronic Commerce, 16*(2), 69–90.

Lin, T.-C., Cheng, H. K., Wang, F.-S., & Chang, K.-J. (2012). A study of online auction sellers' intention to switch platform: The case of Yahoo!Kimo versus Ruten_eBay. *Decision Sciences, 43*(2), 241–272.

Liou, D.-K., Chih, W.-H., Hsu, L.-C., & Huang, C.-Y. (2016). Investigating information sharing behavior: The mediating roles of the desire to share information in virtual communities. *Information Systems and e-Business Management, 14*(2), 187–216.

Lu, J., Wang, L., & Hayes, L. A. (2012). How do Technology readiness, platform functionality and trust influence C2C user satisfaction? *Journal of Electronic Commerce Research, 13*(1), 50–69.

Luis Osorio, A., Afsarmanesh, H., & Camarinha-Matos, L. M. (2011). A service integration platform for collaborative networks. *Studies in Informatics and Control, 20*(1), 19–30.

Luo, X., Li, H., Zhang, J., & Shim, J. P. (2010). Examining multi-dimensional trust and multi-faceted risk in initial acceptance of emerging technologies: An empirical study of mobile banking services. *Decision Support Systems, 49*(2), 222–234.

Lusch, R. F., & Nambisan, S. (2015). Service innovation: A service-dominant logic perspective. *MIS Quarterly, 39*(1), 155–175.

Madhusudan, T., & Uttamsingh, N. (2006). A declarative approach to composing web services in dynamic environments. *Decision Support Systems, 41*(2), 325–357.

Madni, A. M. (2012). Adaptable platform-based engineering: Key enablers and outlook for the future. *Systems Engineering, 15*(1), 95–107.

Magnini, V. P. (2011). The implications of company-sponsored messages disguised as word-of-mouth. *Journal of Services Marketing, 25*(42128), 243–251.

Mahr, D., Kalogeras, N., & Odekerken-Schroder, G. (2013). A service science approach for improving healthy food experiences. *Journal of Service Management, 24*(4), 435–471.

Majchrzak, A., & Malhotra, A. (2013). Towards an information systems perspective and research agenda on crowdsourcing for innovation. *Journal of Strategic Information Systems, 22*(4), 257–268.

Mikalef, P., Giannakos, M., & Pateli, A. (2013). Shopping and word-of-mouth intentions on social media. *Journal of Theoretical and Applied Electronic Commerce Research, 8*(1), 17–34.

Moon, S. K., Shu, J., Simpson, T. W., & Kumara, S. R. T. (2011). A module-based service model for mass customization: Service family design. *IIE Transactions, 43*(3), 153–163.

Mukhopadhyay, S., de Reuver, M., & Bouwman, H. (2016). Effectiveness of control mechanisms in mobile platform ecosystem. *Telematics and Informatics, 33*(3), 848–859.

Ng, I., Scharf, K., Pogrebna, G., & Maull, R. (2015). Contextual variety, Internet-of-Things and the choice of tailoring over platform: Mass customisation strategy in supply chain management. *International Journal of Production Economics, 159*(0), 76–87.

Nielsen, P., & Aanestad, M. (2006). Control devolution as information infrastructure design strategy: A case study of a content service platform for mobile phones in Norway. *Journal of Information Technology, 21*(3), 185–194.

Ogilvie, T. (2015). How to thrive in the era of collaborative services entrepreneurship. *Research-Technology Management, 58*(5), 24–33.

Ordanini, A., Miceli, L., Pizzetti, M., & Parasuraman, A. (2011). Crowd-funding: Transforming customers into investors through innovative service platforms. *Journal of Service Management, 22*(4), 443–470.

Palka, W., Pousttchi, K., & Wiedemann, D. G. (2009). Mobile word-of-mouth – A grounded theory of mobile viral marketing. *Journal of Information Technology, 24*(2), 122–185.

Palka, W., Pousttchi, K., & Wiedemann, D. G. (2009). Mobile word-of-mouth – A grounded theory of mobile viral marketing. *Journal of Information Technology, 24*(2), 172–185.

Parker, G. G., Van Alstyne, M. W., & Choudary, S. P. (2016). Platform revolution: How networked markets are transforming the economy and how to make them work for you. WW Norton & Company. Chicago

Pekkarinen, S., & Ulkuniemi, P. (2008). Modularity in developing business services by platform approach. *International Journal of Logistics Management, 19*(1), 84–103.
Pera, R., & Viglia, G. (2015). Turning ideas into products: Subjective well-being in co-creation. *Service Industries Journal, 35*(42954), 388–402.
Pon, B., Seppala, T., & Kenney, M. (2014). Android and the demise of operating system-based power: Firm strategy and platform control in the post-PC world. *Telecommunications Policy, 38*(11), 979–991.
Pramatari, K., Evgeniou, T. (Theos), & Doukidis, G. (2009). Implementation of collaborative e-supply-chain initiatives: an initial challenging and final success case from grocery retailing. *Journal of Information Technology, 24*(3), 269–281.
Raivio, Y., & Luukkainen, S. (2011). Mobile networks as a two-sided platform – Case open telco. *Journal of Theoretical and Applied Electronic Commerce Research, 6*(2), 77–89.
Ranerup, A., Henriksen, H. Z., & Hedman, J. (2016). An analysis of business models in Public Service Platforms. *Government Information Quarterly, 33*(1), 6–14.
Rayna, T., Striukova, L., & Darlington, J. (2015). Co-creation and user innovation: The role of online 3D printing platforms. *Journal of Engineering and Technology Management, 37*(0), 90–102.
Reinartz, W., & Ulaga, W. (2008). How to sell services more profitably. *Harvard Business Review, 86*(5), 90.
Reisinger, M., Ressner, L., & Schmidtke, R. (2009). Two-sided markets with pecuniary and participation externalities. *Journal of Industrial Economics, 57*(1), 32–57.
Ritala, P., Golnam, A., & Wegmann, A. (2014). Coopetition-based business models: The case of Amazon.com. *Industrial Marketing Management, 43*(2), 236–249.
Roson, R. (2005). Auctions in a two-sided network: The market for meal voucher services. *Networks & Spatial Economics, 5*(4), 339–350.
Rui, H., & Whinston, A. (2012). Information or attention? An empirical study of user contribution on Twitter. *Information Systems and e-Business Management, 10*(3), 309–324.
Ryu, M., Kim, J., & Yun, J. (2015). Integrated semantics service platform for the internet of things: A case study of a smart office. *Sensors, 15*(1), 2137–2160.
Saini, A., & Johnson, J. L. (2005). Organizational capabilities in e-commerce: An empirical investigation of e-brokerage service providers. *Journal of the Academy of Marketing Science, 33*(3), 360–375.
Schiff, A. (2003). Open and closed systems of two-sided networks. *Information Economics and Policy, 15*(4), 425–442.
Seamans, R., & Zhu, F. (2014). Responses to entry in multi-sided markets: The impact of craigslist on local newspapers. *Management Science, 60*(2), 476–493.
Selander, L., Henfridsson, O., & Svahn, F. (2013). Capability search and redeem across digital ecosystems. *Journal of Information Technology, 28*(3), 183–197.
Settanni, E., Newnes, L. B., Thenent, N. E., Parry, G., & Goh, Y. M. (2014). A through-life costing methodology for use in product-service-systems. *International Journal of Production Economics, 153*(0), 161–177.
Shang, S. S. C., Li, E. Y., Wu, Y.-L., & Hou, O. C. L. (2011). Understanding web 2.0 service models: A knowledge-creating perspective. *Information Management, 48*(42128), 178–184.
Shapiro, C., & Varian, H. R. (1998). *Information rules: A strategic guide to the network economy.* Boston: Harvard Business Press.
Shin, D., Song, J. H., & Biswas, A. (2014). Electronic word-of-mouth (eWOM) generation in new media platforms: The role of regulatory focus and collective dissonance. *Marketing Letters, 25*(2), 153–165.
Song, J. H., & Hollenbeck, C. R. (2015). The value of social presence in mobile communications. *Service Industries Journal, 35*(42349), 611–632.
Spring, M., & Araujo, L. (2013). Beyond the service factory: Service innovation in manufacturing supply networks. *Industrial Marketing Management, 42*(1), 59–70.
Suarez, F. F., & Cusumano, M. A. (2010). The role of services in platform markets. In *Platforms, markets and innovation* (pp. 77–98). Cheltenham: Edward Elgar Publishing.

Suarez, F. F., & Kirtley, J. (2012). Dethroning an established platform. *MIT Sloan Management Review, 53*(4), 35–41.

Tan, B., Pan, S. L., Lu, X., & Huang, L. (2015). The role of IS capabilities in the development of multi-sided platforms: the digital ecosystem strategy of Alibaba. com. *Journal of the Association for Information Systems, 16*(4), 248.

Tay, C. K., & Chen, S. L. (2016). Cost estimation of a service family based on modularity. *International Journal of Production Research, 54*(10), 3059–3079.

Tee, R., & Gawer, A. (2009). Industry architecture as a determinant of successful platform strategies: A case study of the i-mode mobile Internet service. *European Management Review, 6*(4), 217–232.

Thomas, L. D., Autio, E., & Gann, D. M. (2014). Architectural leverage: Putting platforms in context. *The Academy of Management Perspectives, 28*(2), 198–219.

Tranfield, D., Denyer, D., & Smart, P. (2003). Towards a methodology for developing evidence-informed management knowledge by means of systematic review. *British Journal of Management, 14*, 207–222.

Tsakonas, G., & Papatheodorou, C. (2006). Analysing and evaluating usefulness and usability in electronic information services. *Journal of Information Science, 32*(5), 400–419.

Tuunanen, T., & Cassab, H. (2011). Service process modularization: Reuse versus variation in service extensions. *Journal of Service Research, 14*(3), 340–354.

Tyagi, A. K., & Senthil, V. (2015). Library automation in India: assessment of library services platforms. *DESIDOC Journal of Library & Information Technology, 35*(6), 408–416.

Ulrich, K. (1995). The role of product architecture in the manufacturing firm. *Research Policy, 24*(3), 419–440.

Vargo, S. L., & Lusch, R. F. (2008). Service-dominant logic: Continuing the evolution. *Journal of the Academy of Marketing Science, 36*(1), 1–10.

Velu, C. (2015). Business model innovation and third-party alliance on the survival of new firms. *Technovation, 35*(0), 1–11.

Voss, C., & Hsuan, J. (2011). Service science: The opportunity to re-think what we know about service design. *Science of Service Systems, 0*(0), 231–244.

Wang, Y., Yuan, Y., Turel, O., & Tu, Z. (2015). Understanding the development and diffusion of mobile commerce technologies in China: A biographical study with An Actor-Network Theory perspective. *International Journal of Electronic Commerce, 19*(4), 47–76.

Weiss, M., & Gangadharan, G. R. (2010). Modeling the mashup ecosystem: Structure and growth. *R & D Management, 40*(1), 40–49.

World Bank. (2014). *World development indicator*. http://wdi.worldbank.org/table/4.2

Xu, Q., Schmierbach, M., Bellur, S., Ash, E., Oeldorf-Hirsch, A., & Kegerise, A. (2012). The effects of "friend" characteristics on evaluations of an activist group in a social networking context. *Mass Communication and Society, 15*(3), 432–453.

Yaraghi, N., Du, A. Y., Sharman, R., Gopal, R. D., & Ramesh, R. (2015). Health information exchange as a multisided platform: Adoption, usage, and practice involvement in service co-production. *Information Systems Research, 26*(1), 1–18.

Yoo, Y., Boland, R. J., Jr., Lyytinen, K., & Majchrzak, A. (2012). Organizing for innovation in the digitized world. *Organization Science, 23*(5), 1398–1408.

Yuan, S.-T., & Fei, Y.-L. (2010). A synthesis of semantic social network and attraction theory for innovating community-based e-service. *Expert Systems with Applications, 37*(5), 3588–3597.

Yum, H., Lee, B., & Chae, M. (2012). From the wisdom of crowds to my own judgment in micro-finance through online peer-to-peer lending platforms. *Electronic Commerce Research and Applications, 11*(5), 469–483.

Zha, Y., Zhang, J., Yue, X., & Hua, Z. (2015). Service supply chain coordination with platform effort-induced demand. *Annals of Operations Research, 235*(1), 785–806.

Zhang, K. Z. K., Cheung, C. M. K., & Lee, M. K. O. (2012). Online service switching behavior: The case of blog service providers. *Journal of Electronic Commerce Research, 13*(3), 184–197.

Zhang, J., Zha, Y., Yue, X., & Hua, Z. (2016a). Dominance, bargaining power and service platform performance. *Journal of the Operational Research Society, 67*(2), 312–324.

Zhang, M., Shen, X., Zhu, M., & Yang, J. (2016b). Which platform should I choose? Factors influencing consumers' channel transfer intention from web-based to mobile library service. *Library Hi Tech, 34*(1), 2–20.

Zhu, Z., Zhao, J., Tang, X., & Zhang, Y. (2015). Leveraging e-business process for business value: A layered structure perspective. *Information & Management, 52*(6), 679–691.

Zoric, J. (2011). Connecting business models with service platform designs: Exploiting potential of scenario modeling. *Telematics and Informatics, 28*(1), 40–54.

Zhang, [illegible] mobile power and service platform [illegible]
[illegible]
Zhao, [illegible] What [illegible] should [illegible]
[illegible]
[illegible]
[illegible]
[illegible]
[illegible]

Platform Ecosystem Orchestration for Efficiency, Development, and Innovation

Anssi Smedlund, Hoda Faghankhani, Heini Ikävalko, and Petra Turkama

Abstract Platform research has expanded its focus from management of technology domains towards the service of a business. Digital service platforms facilitate ecosystems of participants and compete against each other. Platform ecosystems cannot be managed in a goal-oriented fashion because the number of actors, transactions, and relationships increases beyond the ability of what the platform owner can handle. Instead, platform ecosystems can be orchestrated by designing processes taking place among participants. In this conceptual paper, we present four service platform categories and three platform ecosystem orchestration modes for platform owners. In conclusion, we suggest that platform ecosystems around the service platform require simultaneous orchestration of efficiency, development, and innovation in order to attract and lock-in end users, facilitate transactions, and create novel offerings.

Keywords Platform economy · Service platform · Orchestration

Introduction

Companies owning digital service platforms have become the biggest brands and largest corporations in the world over product brands (Desjardins 2017). A digital service platform allows client firms to build their business on top of service modules offered online (Chesbrough 2011). For end users, service platforms integrate as parts of daily life, for example, in shopping, travel, and rental transactions (Smedlund 2012). The service platforms, such as accommodation service Airbnb and transportation service Uber, are easily scalable and aim to create a multisided market

A. Smedlund (✉) · H. Ikävalko · P. Turkama
Center for Knowledge and Innovation Research (CKIR), Aalto University School of Business, Helsinki, Finland
e-mail: anssi.smedlund@aalto.fi; heini.ikavalko@aalto.fi; petra.turkama@aalto.fi

H. Faghankhani
Cafe Bazaar Co., Tehran, Iran

© Springer Nature Singapore Pte Ltd. 2018

A. Smedlund et al. (eds.), *Collaborative Value Co-creation in the Platform Economy*, Translational Systems Sciences 11,
https://doi.org/10.1007/978-981-10-8956-5_2

resulting in a platform ecosystem. These types of platforms transform the structures of their respective businesses and drive incumbents out of business as participants join in the platform instead of continuing business in a traditional way (Davies et al. 2005). Service platforms are transforming the way products and services are being consumed and are radically changing economic and social structures. Despite their significance, value-creating processes and orchestration are relatively new research topics.

One important aspect in the platform literature is the variety of characteristics included in the service platform discussion. Depending on the number of participating groups, platforms facilitate one-sided, two-sided, or multisided markets (Evans et al. 2005). And, depending on whether participants are free to integrate their offering into the platform without the platform owner specifically choosing the participant, the platform can be either closed or open (Chesbrough 2011). The common denominator to the kinds of platforms discussed in this chapter (i.e. digital service platforms) is that they all facilitate a multisided market around them allowing the ecosystem to grow with network effects and have invested heavily in information and communication technology (ICT) infrastructure. The ICT allows both flexible front-end interactions with each end user individually (e.g. Amazon.com personalized landing page) and also digital service modules for complementors to base their offering on (e.g. Amazon.com billing system for third-party merchants).

The world's most successful service platforms orchestrate their ecosystems of complementors and end users. These platforms have grown evolutionarily, and expanded horizontally to adjacent businesses, thus connecting participants from different industries. They are bridging traditional industry divides and offer highly desirable novel offerings emerging in between the industries. The platform ecosystems include the platform owner, complementors, and end users (McIntyre and Srinivasan 2016) and in the cases of horizontally integrated service platforms, also competitors that may form secondary or complementary platforms and ecosystems (Hänninen et al. 2017; Thomas et al. 2014).

So far, the literature on platforms lacks theories and empirical studies on platform ecosystem orchestration. In innovation management, orchestration is defined as facilitating the processes that lead to and promote relationships and activities among the participants (Dhanaraj and Parkhe 2006; Teece 2007). In the case of service platforms, orchestration can influence how well actors in the ecosystem innovate novel offerings, develop them incrementally, and organize to execute the transactions.

The purpose of this chapter is to highlight the differences of service platforms and the need for platform ecosystem orchestration. In this paper, we conceptualize four kinds of service platforms and articulate the distinct needs for orchestration of each platform type. Then we present three platform ecosystem orchestration modes the platform owner can pursue in order to improve the efficiency and develop and renew business among participants. It is concluded that successful platforms should engage in orchestration not only for efficiency but also for development and innovation. Our theorizing reveals that each of the orchestration modes is different in terms of goals, value co-creation logic, interdependencies, sources of synergy, and growth patterns. This provides a starting point for designing particular orchestration processes in platform ecosystems.

Platform Ecosystems and Orchestrator Role

Based on a review of the platform literature, we conclude that platform ecosystems have four interrelated commonalities: (1) co-creation of value, (2) interdependency and complementarity of participants, (3) synergy, and (4) evolutionary growth. Firstly, the end user value is a result of value constellation, as several firms and also the end users themselves participate in value co-creation (Ceccagnoli et al. 2012; Normann and Ramirez 1993). Secondly, platform ecosystem participants are in complementary relationship with each other, which is necessary to ensure functionality of the entire offering (Gawer and Cusumano 2008; Gawer and Henderson 2007; Nishino et al. 2012). Thirdly, the platform ecosystem produces surplus value (i.e. synergy) as a result of complementarity and interdependency of components – the value co-created in aggregation of each component as a whole system is more than the sum of values created by each component separately (Armstrong 2006).

Fourthly, platform ecosystems adapt to their environment: they expand by either building upon new components or connecting to other ecosystems. After reaching a tipping point of momentum in the number of participants and relationships between them, ecosystems develop in an evolutionary manner (i.e. random variation, selection, and retention processes). An evolutionary attribute is necessary because it allows the platform to maintain its current participants and simultaneously attract new ones.

The platform owner is the focal point of the platform ecosystem (McIntyre and Srinivasan 2016). Empirical studies of precisely how the platform owner fosters the emergence of the ecosystem are non-existing, but conceptually there are two processes on which ecosystems are formed: goal-directedness and serendipity. In a goal-directed process, the participants see themselves as a part of a network committed to some common goal. The ecosystem is formed to achieve this goal. In the serendipitous process, there is no pre-existing goal, and the network develops in an evolutionary manner (Kilduff and Tsai 2003).

For example, the process by which Apple orchestrated its iPod product and iTunes service is an example of how platform growth changes from a goal-oriented to a serendipitous process – from executing a focal firm's strategy into an inter-industry-wide ecosystem. In the beginning, Apple developed the iPod in collaboration with selected companies: a couple of technology providers and one content provider. A goal was set to bring a new music player to the market. In 2004, the network was expanded with tens of content producers, technology producers, and also producers of peripheral devices. It is notable that several of Apple's competitors joined the network, intermediated by the technology providers.

As a result of serendipitous networking, there are many participants who may have differing business goals with Apple but are complementary to the platform ecosystem. This leads to a decentralized and distributed network structure, as the offerings of multiple firms in the ecosystem are compatible. The network spreads out to include a wide variety of businesses, and the platform ecosystem starts to connect different domains of industries together.

The Apple iPod case illustrates how an ecosystem emerges around the platform owner with a combination of goal-oriented and evolutionary growth. The ecosystem is comprised not only of all the firms that belong to the immediate value chain with Apple but also of firms and other stakeholders that somehow affect the value chain (Iansiti and Levien 2004). Each of the actors in the ecosystem has its own niche that is connected to the ecosystem with complementary products or services.

The platform literature emphasizes that the platform owner is in the position of orchestrating the platform ecosystem. The orchestrator strives to share the standards (Gueguen and Isckia 2011), develops the industry vision (Gueguen and Isckia 2011; Moore 2006, 1996), maintains the integrity of the platform and its evolution (Gawer and Cusumano 2002; Gawer and Henderson 2007), and determines up to a point who may join as a participant in the ecosystem (Eisenmann 2008). The orchestrator role has been referred to by different terms in different studies. Some studies have used 'platform leader' (Cusumano and Gawer 2002; Gawer and Cusumano 2008; Gawer and Henderson 2007; Moore 1996; Tee and Gawer 2009), some have used the term 'platform sponsor' (Alstyne et al. 2011; Eisenmann 2008; Eisenmann et al. 2008; Parker and Van Alstyne 2008), and others have used 'keystone organizations' (Gueguen and Isckia 2011; Hanssen 2012; Iansiti and Levien 2004).

Service Platform Ecosystem Orchestration for Efficiency, Development, and Innovation

In this section, we conceptualize different kinds of service platforms and explain their particular needs for platform ecosystem orchestration. We argue that service platforms differ in terms of focus of ICT investments and openness of the platform for participants. Based on these distinguishing factors and inspired by Gawer and Cusumano (2014), we have labelled the resulting platforms as internal platforms, industry-wide open or closed platforms, and inter-industry platforms.

ICT acts as a multiplier of value in the platform ecosystems because it affects the possibilities of platform participants to co-create value within sides and between sides. Digitalization helps diverse actors collaborate in the ecosystem and support the constant evolution of the actors in the ecosystem (Lusch and Nambisan 2013). In addition to the technological core, ICT investments contribute directly to the applicability of accumulated data used, back-end processes, service modules and boundary resources offered for complementors, and user experience of a flexible front end. A focus on ICT investments, especially in either back-end or front-end ICT, distinguishes platforms from one another (Smedlund 2012). Seemingly, there are service platform owners that consider user experience (UX) of the front-end as a differentiator from other platforms (e.g. Rakuten), while others focus on back-end processes and supplier modules (e.g. Alibaba) (Hänninen et al. 2017).

Another distinguishing factor is open or restricted entry to the platform. Openness goes hand in hand with the assumption of increased amount of participating groups

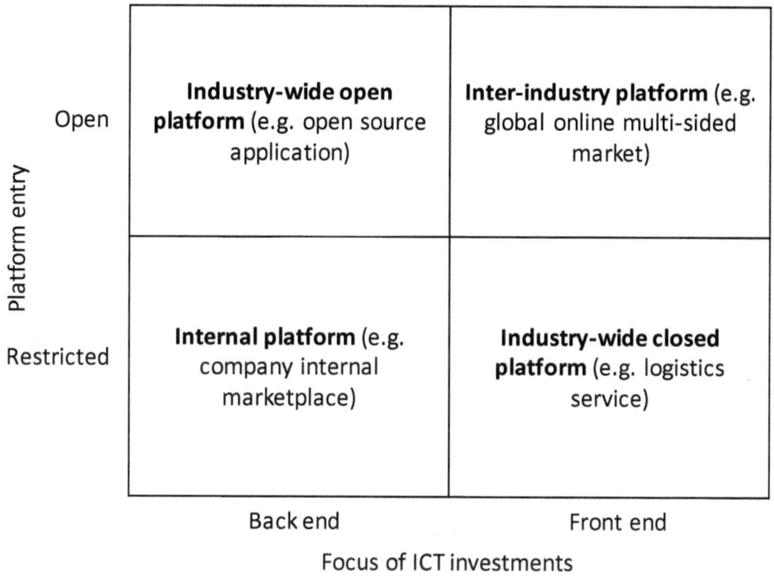

Fig. 1 Four types of service platforms

and segments. If the industry-wide platform is open, then the amount of collaboration (e.g. negotiations, integration, flow of resources and money) between the participants in the platform is likely to be high (Smedlund 2012). When these increase, the complexity of the network of relationships in the platform ecosystem increases. If the platform entry for participants is restricted, the relationship between the platform owner and participants, and among participants, is likely to be more formalized, which lowers the need for communication and coordination. Figure 1 presents the four types of platforms.

The type of service platform determines contingencies on what the platform owner should focus orchestration on. The three modes of orchestration, efficiency, development, and innovation, describe how a platform owner can enhance the ecosystem from different angles. The three modes have roots in management theory, organization theory, and complex systems theories.

In management theory, efficiency is highlighted in theories of scale and scope of production (Chandler 1962), development from the point of view of theories of knowledge accumulation (Nelson and Winter 1982; Penrose 1959), and innovation from the point of view of technological change and entrepreneurship as forces transforming society (Schumpeter 1934, 1942; Smedlund 2009a, b). In organization theories, a rational system point of view denotes efficiency in organizing to achieve a predetermined goal, natural system point of view brings in incremental organizational development in order to adapt to many goals simultaneously, and open system point of view highlights the organization's ability to self-organize in accordance with the environment (Scott 2003). In complex systems theories, the three modes of orchestration are present in the paradigms of systems thinking (Ståhle et al. 2003):

efficiency in the mechanistic Newtonian and classical physics paradigm, development in general systems theory (Bertalanffy 1969) and soft systems methodology (Checkland and Scholes 1990), and innovation in self-organizing systems (Prigogine 1976) and in autopoietic systems (Maturana and Varela 1980).

Next, each of the orchestration modes is connected to a platform type, and interrelations between (1) co-creation of value, (2) interdependency and complementarity of components, (3) synergy, and (4) evolutionary growth pattern are theorized. After this, concrete orchestration processes can be suggested.

In internal platforms (e.g. company internal marketplace in a multinational corporation), the participants are well known and specified, making collaboration predictable and ICT systems workable with minimum requirements. Flexible front-end UX is not a top priority when the participants are known because organizational boundaries restrict participation making the payoff for any additional investments in front-end UX low.

This type of platform ecosystem requires orchestration focusing on efficiency. The goal is predetermined as in a rational system (Scott 2003) to make transactions as efficient as possible – the more transactions there are, the more efficient the platform should become. This makes possible synergies from bundling complementor offerings by lowering the transaction costs of end users. The interdependencies among participants are transaction based. The value co-creation logic is Newtonian mechanistic (Ståhle et al. 2003), machine-like, and is based on deals between participants that take place in the platform. In an internal platform, the offering and the value constellation are known, and value is co-produced with a known set of network actors (c.f. Ramirez 1999). Concrete orchestration processes for efficiency are, for example, facilitation of transactions, managing reliability of participants, and transactions and processes that make efficient use of existing explicit knowledge possible (Smedlund 2009a, b).

In closed industry-wide platforms (e.g. logistics service provider's web service for tracking parcels), the participants are also well known because the platform owner decides whether the participant can join the platform or not. There are only selected external participants who are related to the business of the platform owner (e.g. in the case of parcel delivery tracking system), but they do not belong to the same organization. The participants' attention is directed towards the specific transaction making requirements for ICT straightforward and predictable. Front-end ICT is important in closed platforms, as the differing end user needs have to be served well with the combination of the capabilities and assets of the participants. Front-end design enables end users serving themselves instead of the platform owner having to target customer service resources for end users.

The participant of an industry-wide open platform (e.g. end user of a snippet of code for software in open source code repository) may be not specified at all but is likely to belong to a predefined group. This is because the participants are more bound towards collaboration in a specific substance area – the participants are looking for some existing solutions to complement their own offering or add their solution to an existing offering. Here the platform owner does not restrict participants' options of joining the platform but does not invest heavily in front-end ICT either.

The participants assemble their own user-specific combinations of the ecosystem offerings by using a front-end ICT system that requires skills and dedication to use, but is also flexible to serve different needs.

Open and closed industry platforms are depicted here as facing similar kinds of orchestration modes. In an industry-wide open or closed platform, interactions among or between participants require more tailoring than in an internal platform. In industry-wide platforms, the participants are more or less predictable, but do not reside inside a shared organizational boundary. In other words, many other goals in addition to the goals of the platform owner exist simultaneously, similar to organization as a natural system (Scott 2003). Knowledge is more complex compared to the internal platform because of the added variety of participants. This affects the value co-creation logic making it organic with ongoing dialogue, soft systems like Ståhle et al. (2003). An industry-wide platform is a venue for co-learning (Ramirez 1999) and co-elevation (Kijima and Arai 2016).

In the industry-wide open or closed platform ecosystem, the participants' capabilities are continuously improved. The management of interdependencies should be directed towards facilitating reciprocal interactions among participants. Sources of synergy come from co-learning and co-elevation of capabilities of participants. The offering is incrementally improved thus making the offering more than just a bundle of existing modules. In these platforms participants gain tacit, experience-based knowledge (Smedlund 2009a, b). Thus industry-wide platforms require orchestration of processes to facilitate long-term reciprocal interactions and processes of retention and refining experience-based knowledge. When engaged in such activities, the participants are being locked into the platform, which leads to increased robustness.

Inter-industry platforms (e.g. global online multi-sided market) attract and connect a wide variety of participants thus setting high requirements for both collaboration and ICT. Because openness leads to increased variety, inter-industry platform ecosystems are highly complex compared to other types. UX on the front end is crucial; otherwise the varieties of participants are not able to connect themselves fully to the platform, but so are the back-end processes and handling of accumulated participant-specific data. Orchestration in inter-industry platforms should be focused on creating novel offerings with platform participants. The value co-creation logic resembles the logic of self-organizing (Prigogine 1976), and novel offerings emerge in the boundaries between different industrial domains. Novel value constellations are co-invented in a loosely coupled interaction between the participants (Ramirez 1999). The competencies are created and renewed according to feedback from the ecosystem, as in organizations functioning with open systems logic (Scott 2003). In inter-industry platform, the orchestrator should facilitate processes that result in co-experiencing in ad hoc interactions (Kijima and Arai 2016), emergence of new offerings, and creation of new knowledge. These activities lead to increased diversity and improved reach across industrial domains. Table 1 summarizes orchestration for efficiency, development, and innovation and platform types they are most suitable for.

Table 1 Orchestration of internal, industry-wide, and inter-industry platforms

Orchestration mode	Efficiency	Development	Innovation
Platform type most suitable	Internal platform	Industry-wide platform (open/closed)	Inter-industry platform
Value co-creation logic	Mechanistic, supplier-buyer deals. Value co-production	Organic, ongoing dialogue. Value co-learning and co-elevation	Chaotic, self-organizing. Value co-invention and co-experiencing
Interdependencies	Contractual, transaction-based	Reciprocal	Trust-based
Sources of synergy	Lowering transaction costs, bundling offerings	Incremental improvements, co-elevation of capabilities	Novel offerings
Growth pattern	Repeated transactions leads to platform efficiency	Retention of participants leads to platform robustness	Innovation activities lead to more platform diversity and reach
Orchestration processes (e.g.)	Facilitate transactions, manage reliability, use of existing knowledge	Facilitate long-term reciprocal interactions, retention and refining of knowledge	Emergence of new offerings, facilitate ad hoc interactions, creation of new knowledge

Discussion

Digital service platforms are changing structures of our society and economy. Service platforms are orchestrators of platform ecosystems, and the ecosystems are interconnected as participants contribute to several ecosystems. Platform ecosystems are competing against each other for end users and participants, and today's platform leaders may be quickly dethroned tomorrow (McIntyre and Srinivasan 2016; Suarez and Kirtley 2012). To attract and lock-in end users, platform owner's ability to renew the offering is essential in competition. As soon as the ecosystem stops evolving, the offering becomes static, and the platform becomes a target of envelopment (Tiwana 2014). This distinguished, for example, Myspace from Facebook in the social media industry and Napster from Spotify in the music industry. However, also some core elements are needed for participants to stick with the platform owner: accumulated historical user data and operational efficiency are important in attracting and locking-in end users.

The platform literature up until now is technology oriented and does not fully explain the platform ecosystem phenomenon. Until recently, platforms have been mainly approached from the management of technology point of view, and platform strategies have been studied mainly from the point of view of managing the supply side of the platform (Lusch and Nambisan 2013). In the digital service platforms, many of the value co-creation processes are of a social kind and provide intangible value and experiences for end users instead of technological performance and clearly measurable value for suppliers.

Based on existing definitions of platforms (Gawer and Cusumano 2014), we presented four platform categories, out of which the inter-industry platform, being open for participants to enter and focusing the ICT investments towards front end in addition to back end, is typical for those platforms that integrate horizontally across industry divides and eventually become global successes. One conclusion of this chapter is that the inter-industry platform type should be treated as a separate category in addition to internal and industry-wide open or closed platform categories. It was stated that internal platforms should be orchestrated for efficiency, industry platforms for development, and inter-industry platforms for innovation.

Successful platforms should engage in orchestration processes of all these modes simultaneously in order to renew their offering and maintain their position in the market. Novel offerings are unlikely to emerge without a high amount of inter-participant collaboration that requires a superior front end for the digital platform. Similarly, highly efficient transactions require a known set of participants and attention to back-end processes. Platform system architecture should include elements of all four above-mentioned platform types, internal, industry-wide open and closed, and inter-industry platform, and it should simultaneously orchestrate processes that aim for efficiency, development, and innovation. Figure 2 presents a conceptual model of the platform orchestration loop.

The idea of the platform orchestration loop is that different orchestration processes result as a self-enforcing loop leading to platform evolution and growth. The loop goes from loose coupling between the platform participants to tight coupling and from value constellation creation to executing the value network (Lusch and Nambisan 2013; Ramirez 1999). Novel offerings for end users emerge or, from the point of view of an outside observer, self-organize in the interfaces between industry domains. These novel offerings attract end users to the platform, and as the end users consume the offerings, transactions are facilitated. Efficient transactions further lock-in end users. As end users are being locked-in, the feedback from the users is likely to set ground for the next round of platform renewal as novel offerings emerge again. In a platform capable of sustaining its market position over time, value constellation that is co-produced in the efficiency orchestration mode is combined and co-elevated in the development orchestration mode and co-invented in the innovation orchestration mode (c.f. Ramirez 1999).

Platform entry affects the amount of ad hoc collaboration between the participants. As ad hoc collaboration activity increases, the complexity of knowledge is also increased. Novel offerings and related value constellation emerge when these two factors are present. Clever investments in ICT supporting this and innovation orchestration result in the highest potential synergies. As soon as the value constellation organizes itself and the end users are attracted to co-produce it, the operating mode of the platform changes to the mode of efficiency, and executing the transaction becomes more or less routine activity for the participants of value constellation. In this case, ICT investments in back-end processes and efficiency orchestration create steady income for the platform.

In this chapter, we theorized four service platform types and three platform ecosystem orchestration modes. It can be concluded that the most successful digital

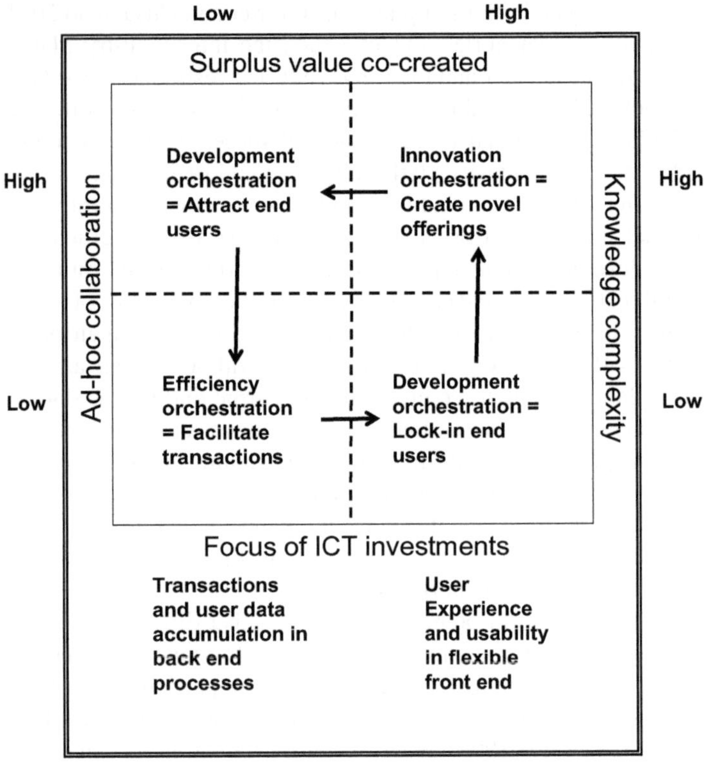

Fig. 2 Platform orchestration loop

service platforms are ambidextrous: they not only manage to empower third parties to innovate novel offerings but also pay attention to efficiency of transactions and incremental development.

Acknowledgement This research has received financial support from the European Union Horizon 2020 Programme under grant agreement no. 688203.

References

Alstyne, V., Eisenmann, T., Parker, G., & Van Alstyne, M. (2011). Platform envelopment. *Strategic Management Journal, 32*(May 2007), 1270–1285.

Armstrong, M. (2006). Competition in two-sided markets. *The Rand Journal of Economics*. http://onlinelibrary.wiley.com/doi/10.1111/j.1756-2171.2006.tb00037.x/abstract

Bertalanffy, L. V. (1969). *General system theory; Foundations, development, applications*. New York: G. Braziller.

Ceccagnoli, M., Forman, C., Huang, P., & Wu, D. J. (2012). Cocreation of value in a platform ecosystem: The case of enterprise software. *MIS Quarterly, 36*(1), 263–290.

Chandler, A. D. (1962). *Strategy and structure: Chapters in the history of the industrial enterprise*. Cambridge: MIT Press.

Checkland, P., & Scholes, J. (1990). *Soft systems methodology in action*. New York: Wiley.

Chesbrough, H. (2011). *Open services innovation*. San Francisco: Jossey-Brass.

Cusumano, M. A., & Gawer, A. (2002). The elements of platform leadership. *MIT Sloan Management Review, 43*(3), 51–58.

Davies, M., Pitt, L., Shapiro, D., & Watson, R. (2005). Betfair.com: Five technology forces revolutionize worldwide wagering. *European Management Journal, 23*(5), 533–541.

Desjardins, J. (2017). RANKED: The most valuable brands in the world. *Business Insider.com*. http://www.businessinsider.com/most-valuable-brands-ranked?r=US&IR=T&IR=T

Dhanaraj, C., & Parkhe, A. (2006). Orchestrating innovation networks. *Academy of Management Review, 31*(3), 659–669.

Eisenmann, T. R. (2008). Managing proprietary and shared platforms. *California Management Review, 50*(4), 31–54.

Eisenmann, T. R., Parker, G., & Van Alstyne, M. (2008). Opening platforms: How, when and why? *Harvard Business School Working Papers*, 09–030.

Evans, D., Hagiu, A., & Schmalensee, R. (2005). A survey of the economic role of software platforms in computer-based industries. *CESifo Economic Studies, 51*(2–3), 189–224.

Gawer, A., & Cusumano, M. A. (2002). *Platform leadership: How Intel, Microsoft, and Cisco drive industry innovation*. Boston: Harvard Business School Press.

Gawer, A., & Cusumano, M. A. (2008). How companies become platform leaders. *MIT Sloan Management Review, 49*(2), 28–35.

Gawer, A., & Cusumano, M. A. (2014). Industry platforms and ecosystem innovation. *Journal of Product Innovation Management, 31*(3), 417–433.

Gawer, A., & Henderson, R. (2007). Platform owner entry and innovation in complementary markets: Evidence from Intel. *Journal of Economics & Management Strategy, 16*(1), 1–34.

Gueguen, G., & Isckia, T. (2011). The borders of mobile handset ecosystems: Is coopetition inevitable? *Telematics and Informatics, 28*(1), 5–11.

Hänninen, M., Smedlund, A., & Mitronen, L. (2017). Digitalization in retailing: Multi-sided platforms as drivers of industry transformation. *Baltic Journal of Management*, BJM-04-2017-0109.

Hanssen, G. K. (2012). A longitudinal case study of an emerging software ecosystem: Implications for practice and theory. *Journal of Systems and Software, 85*(7), 1455–1466.

Iansiti, M., & Levien, R. (2004). Strategy as ecology. *Harvard Business Review, 82*(3), 68–78.

Kijima, K., & Arai, Y. (2016). Value co-creation process and value orchestration platform. In S. Kwan (Ed.), *Global perspectives on service science: Japan*. New York: Springer.

Kilduff, M., & Tsai, W. (2003). *Social networks and organizations*. London/Thousand Oaks: SAGE.

Lusch, R. F., & Nambisan, S. (2013). Service innovation: A service-dominant (S-D) logic perspective service. *MIS Quarterly*. http://air.eller.arizona.edu/docs/papers/2012/Service_Innovation_Lusch_Nambisan.pdf

Maturana, H., & Varela, F. J. (1980). *Autopoiesis and cognition: The realisation of the living*. London: Reidl.

McIntyre, D. P., & Srinivasan, A. (2016). Networks, platforms, and strategy: Emerging views and next steps. *Strategic Management Journal* (in Press).

Moore, J. F. (1996). The death of competition: Leadership and strategy in the age of business ecosystems. *Harper Business*. http://www.getcited.org/pub/100107305

Moore, J. (2006). Business ecosystems and the view from the firm. *Antitrust Bull.* http://heinonlinebackup.com/hol-cgi-bin/get_pdf.cgi?handle=hein.journals/antibull51&section=9

Nelson, R. R., & Winter, S. G. (1982). *An evolutionary theory of economic change*. Cambridge, MA: Belknap.

Nishino, N., Wang, S., Tsuji, N., Kageyama, K., & Ueda, K. (2012). Categorization and mechanism of platform-type product-service systems in manufacturing. *CIRP Annals – Manufacturing Technology, 61*(1), 391–394.

Normann, R., & Ramirez, R. (1993). From value chain to value constellation: Designing interactive strategy. *Harvard Business Review, 71*(4), 39–51.

Parker, G., & Van Alstyne, M. (2008). Managing platform ecosystems. ICIS 2008 Proceedings. 53. See: http://aisel.aisnet.org/icis2008/53/

Penrose, E. T. (1959). *The theory of the growth of the firm*. New York: Wiley.

Prigogine, I. (1976). Order through fluctuations: Self-organization and social systems. In E. Jantsch & C. H. Waddington (Eds.), *Evolution and eonsciousness: Human systems in transition*. Reading: Addison-Wesley.

Ramirez, R. (1999). Value co-production: Intellectual origins and implications for practice and research. *Strategic Management Journal, 20*, 49–65.

Schumpeter, J. A. (1934). *The theory of economic development: An inquiry into profits, capital, credit, interest and the business cycle* (2nd ed.). Cambridge, MA: Harvard University Press.

Schumpeter, J. A. (1942). *Capitalism, socialism and democracy*. London: Unwin University Books.

Scott, W. R. (2003). *Rational, natural, and open systems* (5th ed.). Upper Saddle River: Prentice Hall.

Smedlund, A. (2009a). *Network approach to fundamental tasks in knowledge-based organizations*. Dr. Sc. dissertation, Helsinki University of Technology, Faculty of Information and Natural Sciences, Department of Industrial Engineering and Management. http://lib.tkk.fi/Diss/2009/

Smedlund, A. (2009b). Social network structures for explicit, tacit and potential knowledge. *International Journal of Knowledge Management, 5*(1), 78–87.

Smedlund, A. (2012). Value cocreation in service platform business models. *Service Science, 4*(1), 79–88.

Ståhle, P., Ståhle, S., & Pöyhönen, A. (2003). *Analyzing dynamic intellectual capital: System-based theory and application* (Acta Universitatis Lappeenrantaensis 152). Lappeenranta: Lappeenranta University of Technology., http://scholar.google.com/scholar?hl=en&btnG=Search&q=intitle:Lappeenrannan+teknillinen+yliopisto#9

Suarez, F. F., & Kirtley, J. (2012). *Dethroning an established platform dethroning an established platform*, 53(53414).

Tee, R., & Gawer, A. (2009). Industry architecture as a determinant of successful platform strategies: A case study of the i-mode mobile internet service. *European Management Review, 6*(4), 217–232.

Teece, D. J. (2007). Explicating dynamic capabilities: The nature and microfoundations of (sustainable) enterprise performance. (N. E. Al, Ed.) *Strategic Management Journal, 1350*(August), 1319–1350.

Thomas, L. D., Autio, E., & Gann, D. (2014). Architectural leverage: Putting platforms in context. *The Academy of Management Perspectives, 28*(2), 198–219.

Tiwana, A. (2014). *Platform ecosystems: Aligning architecture, governance, and strategy*. Waltham: Morgan Kaufmann.

Platform as a Social Contract: An Analytical Framework for Studying Social Dynamics in Online Platforms

Joni Salminen, Nicolas Gach, and Valtteri Kaartemo

Abstract In addition to formal terms of service and contracts between platform owners, users, and other stakeholders, there can be seen an implicit social contract taking place in online platforms, and influencing the social dynamics, such as trust, expectations, and perceived social justice, taking place within platforms, and driving their growth and success in the background. This paper examines the nature of that social contract, to better understand the complex social dynamics taking place in online platforms. To accomplish that objective, we draw from classic Enlightenment thinkers, e.g., Rousseau, Locke, and Hobbes, to analyze key aspects of social contracts, which we define as the alignment of stakeholder interests, stakeholder support, economic and social justice, and transparency of expectations. As our main contribution, we develop a conceptual framework for the analysis of platforms based on social contract theory, the Platforms as a Social Contract framework. The applicability of the framework is illustrated through a case analysis of YouTube, a popular online content platform. The rich understanding provided by the social contract perspective, embodied in our framework, entails many potential advantages to platform owners, including understanding user motivations and reactions so that effective platform governance with maintaining a sustainable solution to the chicken-and-egg problem becomes possible. While individual platforms may come and go, each faces the same fundamental social dynamics that can be explained

J. Salminen
Turku School of Economics, Turku, Finland

Qatar Computing Research Institute, Hamad Bin Khalifa University, Doha, Qatar
e-mail: joni.salminen@utu.fi

N. Gach
Hastor Consulting, Paris, France
e-mail: nicolas.gach@ngc-transform.com

V. Kaartemo (✉)
Turku School of Economics, Turku, Finland
e-mail: valtteri.kaartemo@utu.fi

© Springer Nature Singapore Pte Ltd. 2018

A. Smedlund et al. (eds.), *Collaborative Value Co-creation in the Platform Economy*, Translational Systems Sciences 11,
https://doi.org/10.1007/978-981-10-8956-5_3

and understood by applying the social contract framework presented in this research. This research shows how the framework can be used for analysis of online platforms, as well as suggests future research avenues for developing deeper understanding of platforms as a social contract.

Keywords Online platforms · Social contract · Trust · Two-sided markets · Chicken-and-egg problem

Introduction

> Find a form of association that will bring the whole common force to bear on defending and protecting each associate's person and goods, doing this in such a way that each of them, while uniting himself with all, still obeys only himself and remains as free as before. (Jean-Jacques Rousseau, *Du Contrat Social (Of Social Contract)* (1762), Book 1, Chapter 6)

Motivation

Platforms, defined as places of interaction (Salminen 2014), entities to "build on top of" (Gawer 2011), or two-sided markets[1] with demand- and supply-side actors (Rochet and Tirole 2003), play an important role in modern economy for several reasons. First, economic activity is provided through platforms in large volumes. Second, increasing share of individuals' time and effort is concentrated on activities conducted on platforms. For example, an increasing number of people are driving Uber as an alternative to traditional employment. Consequently, platforms exert considerable economic and social power over their users (Armstrong 2006; Gawer 2011). In this research, we claim that platforms can be understood as structures enabling social contracts and that such understanding will greatly enhance their governance from the platform owner's (i.e., the entity responsible for managing the structure, later referred to as PO) perspective.

As a practical example, consider online advertising; we define it as a form of social contract in which the users give away some parts of their privacy, and accept being shown targeted advertisement, in exchange for free access and use of the platform (see Anderson and Gabszewicz 2006). Even though this exchange is not formally defined as a contract, apart from the terms of service (TOS) of the platform, both the users and POs implicitly consent to it, as evidenced by the millions and millions of interactions carried out in free-to-use online platforms every day. When this implicit agreement is not upheld by either party or in other words when

[1] In addition to two-sided markets, which is a concept rooted in economics, other concepts referring to platforms, although more rarely used, include multisided platforms (Hagiu and Wright 2015) and multisided markets (Seamans and Zhu 2013).

the trust framework surrounding the contract is broken (e.g., too frequent, invasive or misplaced use of advertisement), dissident forms of action occur, such as the use of advertisement content blockers, as an answer to the breach of agreement.

Surprisingly, despite the apparent similarities between platforms as structures (i.e., entities organizing systems of interacting and interdependent elements) and social contract theory, the notion of social contract has not received much attention in the platform context. In fact, we were unable to find any research explicitly discussing platforms from the perspective of social contract. As such, the understanding of trust, exchange, and performance in the context of platforms remains limited. While there are existing works that discuss platforms from transaction cost perspective (e.g., Susarla et al. 2009), social contract theory can provide an alternative understanding to dynamics taking place between POs and stakeholders that go beyond the conventional economic reasoning. We intend to address this gap in current literature by conceptualizing platforms as a social contract (PaaSC).

In particular, we apply the social contract perspective to find similarities to the challenges of modern online platforms. To clarify, we are not explicitly interested in contracts as legal agreements or formal acceptance from the users – rather, we follow the definition that the social contract is an abstraction dealing with ways of how to organize the society or structure (in our case, the platform). Our endeavor is motivated by the premise that an increased understanding of the social reality of platforms leads to discovering effective solutions to various strategic challenges.

For example, for finding sustainable solutions to the well-known and persistent chicken-and-egg problem,[2] taking place in two-sided markets and other platforms (Rochet and Tirole 2003; Caillaud and Jullien 2003; Armstrong 2006), a deep understanding of the prevailing social dynamics taking place in the platform is instrumental, as joining and actively using a platform is ultimately a social phenomenon. By social dynamics, we refer to several constructs, including expectations, trust, stakeholder interests, perceived economic and social justice, and support for the social contract. Such issues are scarcely discussed in the extant platform theory. By analyzing them with social contract theory, we provide additional analytical tools for the POs and scholars interested in platforms. Our objective is therefore to introduce conceptual insights from the rich tradition of social contract theory to economics-dominated platform literature.

The PaaSC framework represents a novel approach for understanding user-related issues of platforms, an area of research which is currently dominated by theories from economics, and in a considerable need for alternative explanations (Birke 2008; Shy 2011; Salminen 2014). By only applying theories from economics, the POs may overlook important nuances as to why the end users and other stakeholders behave the way they do in the "real world" and in the process risk either foregoing opportunities of value appropriation (Salminen 2014) or becoming vulnerable to negative tipping, i.e., accelerating flight of users (Katz and Shapiro 1985). To prevent these adverse effects, our approach can help explain the social

[2] The problem is defined as follows: In the absence of others, users and other stakeholders lack the incentive to join or use the platform (see, e.g., Rochet and Tirole 2003; Salminen 2014).

dynamics after the user/other stakeholder joins a platform; such insight can be applied to improve the platform's viability under the intense competition faced by modern online platforms.

Research Questions

When adopting the perspective of social contract theory to understand modern online platforms, several questions emerge. In this research, to achieve our objective, we consider the following questions:

RQ1: What are the constituents of the social contract in the context of modern online platforms?

RQ2: What rights are the users giving up, in exchange for what benefits?

RQ3: How can the users trust in the good will of the PO as a ruler of the platform?

For our inquiry, it is essential to explore the nature of social contract in the context of platforms. This study focuses on establishing this first step, thereby laying ground work for bridging the social contract tradition with modern online platforms. We do not expect to miraculously compress hundreds of years of literature on this topic into one book chapter; instead, our aim is to show useful analogies and open the underlying social reality of platform with the insight from social contract theory. In particular, our framework is based on a rich foundation of classical thinkers and is applicable to the contemporary context of online platforms when platforms are considered as a set of guiding principles for building a structure, i.e., a way to organize elements of a system. Social contract theory applies to designing, building, and sustaining such a structure, hence its applicability to modern online platforms. By considering this rich tradition, we are able to conceptualize platforms as social contracts and better understand the efficient mechanisms for their governance.

We proceed by defining the concept of social contract, as well as introducing the most prominent thinkers in this field. After that, we will compare their ideas with modern reality by giving some examples and elaborations. Then, by synthesizing some prominent concepts from social contract literature, we will create a coherent framework that utilizes the social contract theories' arguments for the examination of online platforms. In the process, we are bridging the modern technology and economics dominant discussion on platforms to the rich philosophical tradition of social contract theory. Through this juxtaposition, we hope to establish a better way of understanding social dynamics in online platforms. Finally, we will demonstrate the application of the framework through a case example. Conclusion and discussion will follow, also opening future research avenues.

Literature Review

Defining the Social Contract and Its Tradition

Social contract is a central concept in moral and political philosophy. It deals with the relationship of government and the people. Historically, the theory of social contract has played an important role, for example, in the French Revolution and the United States Declaration of Independence. Its conceptual roots can be found in the Age of Enlightenment, although, as stated previously, philosophers have a long tradition of analyzing the role of state and the individual, dating back to Plato and beyond. The key tenet is that individuals give up some of their rights in exchange of the protection for their remaining freedoms. One can distinguish three broad categories of writings to be associated with social contract literature: (1) early political/natural science writings by European philosophers, (2) court rulings which have translated the notion of social contract into laws and legal rights, and (3) contract theory which, in conjunction with the "theory of the firm" (Coase 1937; Williamson 1981), analyzes companies as contractual nodes and proposes a convenient bridge for considering the corporate implications of the notion of social contract.

In this research, we focus on the early political writings by European philosophers from the era of Enlightenment (ca. 1650–1800), a historically important period which laid groundwork for modern Western societies. This decision is taken to correctly understand their original message and therefore the root of social contract theory, rather than focusing on later interpretations of it. However, we do acknowledge that there is an abundance of contemporary and modern political science discussing social contract theory (see, e.g., Oakeshott 1991; Gallopín et al. 2001; Colomer 2002). We proceed by briefly summarizing the ideological legacy of three philosophers, namely, Hobbes, Locke, and Rousseau. For brevity, we focus on these three thinkers who are all considered as highly influential in shaping the social contract theory. The purpose of this literature review is to familiarize the reader with the core idea and conceptual foundation of social contracts. Throughout the paper, we continue referring to other works from the social contract tradition, when applicable to explain particular dynamics pertaining to online platforms.

Enlightenment Philosophers and Social Contract

Hobbes' *Leviathan, or the Matter, Forme and Power of a Commonwealth Ecclesiastical and Civil*, published in 1651, is one of the earliest and most decisive pieces on social contract. Hobbes seeks to give an answer to the following question: Why, although men are created equal, and although this equality (*as in*: identical ability to act and decide for themselves by themselves) is for them a natural right, they relinquish this right and give to other men power to rule over them? The theory on power structures at Hobbes' era invoked God as the justification for political power; merely refusing this idea was not Hobbes' goal, but he also aimed to explain why political power is legitimated by men as a free choice and why they accept it against an apparent natural order of things (i.e., them being equal). According to Hobbes, in a natural state, man fights man due to his fear of dying and to the fact that the first threat to a man's life are other men, an impulse causing war and destruction. In other words, this is not a case of "destruction by chaos," but rather a natural order of destruction. To solve this problem, men abide among themselves to a contract they shape, which terms delegate power – necessarily absolute power writes Hobbes – to other men to rule them. Absolute power means that the ruler is the only one to preserve his natural right, which is even "augmented" by the delegated rights of his fellow men.

Another prominent philosopher, worthy of mentioning in this context, is John Locke, who touched the idea of social contract throughout his career, among others in his work *Two Treatises of Government*, published in 1690. His ideas were influential for the United States Declaration of Independence. Basically, Locke established two fundamental concepts inherent to the notion of social contract but also to – broadly speaking – liberalism: (1) natural law and (2) property rights. The notion of natural law is interesting, as it was manipulated by Dutch philosopher Spinoza into a concept that can be defined as "whatever seems inexorably applicable to the human condition" – such as the law of gravity, for instance. But this concept has been twisted and misused through time, starting with Adam Smith (1994) in his *The Wealth of Nations* presenting the infamous idea of the Invisible Hand as a natural law, a definition which Rifkin (2014) criticizes as being short of a natural law. Rifkin (2014) also proposes the perspicacious remark that if a paradigm is defined *stricto* sensu by a set of natural laws, and the zero-marginal cost society puts in question these natural laws, then one can legitimately say that the root causes and consequences of the rise of platform plays constitute a paradigm change.

Finally, Jean-Jacques Rousseau can be attributed with the concept of social contract. His main work on social contract theory, *Du contrat social ou Principes du Droit Politique* (*Of the Social Contract* or *Principles of Political Right*), was published in 1762. As for Locke, it is worth putting this one in perspective with Rousseau's other writings, as it shapes a coherent whole about Rousseau's take on human nature. Rousseau attempted, similarly to Hobbes, to find out why and how can power become legitimate, even though men are born free and equal. Rousseau established the direct relationship between the people's will (the "general will"), which is the only entity that retains actual power, and its translation into law, which expresses this general will and, in the context of a social contract, is the only sovereign power ruling over this contract. Through a social contract, men relinquish their initial freedom – their natural state – and accept to place their private interests behind the public interests. Rousseau also takes the following stance: the consequence of establishing a social contract is leading men to a political state which is fundamentally better than the natural state they were in.

Theory of Platforms and Two-Sided Markets

In the early 2000s, following the success of such platforms such as Amazon, MySpace (now diminished), and Google, platforms emerged as a contemporary topic of scholarly studies. Entering the vocabulary of economics, two-sided markets (Rochet and Tirole 2003) describe the dynamics of platforms, often from economic analysis and pricing perspective. More recently, it has become apparent that alternative explanations are needed to model and understand the complex social phenomena taking place in platforms (Lampinen et al. 2015; Dillahunt et al. 2016; Lampinen and Cheshire 2016). While concepts in industrial economics and network economics, such as network effects (or externalities), critical mass, and equilibrium of supply and demand side participants (Rochet and Tirole 2003; Armstrong 2006; Parker and Van Alstyne 2005; Hagiu 2006; Evans 2011; Eisenmann et al. 2011; Hagiu and Wright 2014), are effective in describing the overall dynamics taking place in a platform (e.g., lack of success can be explained as a consequence of lack of critical mass of users), they are poor in explaining why, apart from economic rationality, individuals take part in and actively keep using a platform.

For example, the theory of network effects explains this so that participants from one market side derive utility or value from the presence of the other market side (cf. Katz and Shapiro 1985). Intuitively, such an explanation makes sense when observing many real-life platforms, such as dating platforms (e.g., Tinder) where the presence of opposite sex directly determines the usefulness of the application. However, such a coarse explanation, while logically sound, is insufficient in explaining the details as to why one platform flourishes and another perishes. For example, MySpace had a "critical mass" of users and considerable network effects; yet, users rapidly abandoned it *en masse*, shifting to users of alternative social networks, such as Twitter and Facebook. Social contract theory can bring deeper understanding to complement already very established economics-based platform theory.

The proliferation of platforms was quickly observed across many industries, and businesses are increasingly adopted the platform strategy and business to follow the market leaders (Hagiu and Wright 2014) that, according to Eisenmann et al. (2011), source a considerable share of their revenue from platform markets. Overall, the platform theory and concepts developed within the economics domain have spread to several other disciplines, e.g., strategic management (Economides and Katsamakas 2006), information systems sciences (Casey and Töyli 2012; Salminen and Teixeira 2013), and marketing (Sawhney et al. 2005). Yet, the most influential works originate from the domain of economics (see, e.g., Rochet and Tirole 2003; Caillaud and Jullien 2003; Armstrong 2006; Shy 2011, for a review), as the scholars in that domain have been very active in studying platforms and their implications.

According to Salminen (2014), there are several limitations of the economic literature explaining platforms. First, little is known about strategic problems faced by POs beyond the notorious chicken-and-egg problem. Despite some extensions to other strategic issues (e.g., Eisenmann et al. 2011), the chicken-and-egg problem is considered as the fundamental focus of concern in the platform growth and success

(Caillaud and Jullien 2003; Rochet and Tirole 2003; Evans 2011). However, there are several other concerns, e.g., monetization dilemma (Salminen 2014), which represent equally important challenges for POs. Second, the extant platform literature primarily focuses on the "big platforms," such as Amazon, Facebook, and the likes, neglecting start-up companies that have much less maneuvering space for solving the chicken-and-egg problem, e.g., due to small market power (Salminen 2014). Exceptions include Caillaud and Jullien (2003), Evans (2011), and Mas and Radcliffe (2011) who adopt the entrant's perspective to platform markets.

Third, and most importantly in the light of our study, the strategic contribution of the economics literature tends to focus on pricing (Shy 2011; Salminen 2014). The applicability of pricing in the online environment is curbed by "zero pricing," i.e., setting the price level to zero, as a default requirement in the industry (Salminen 2014). For example, Rochet and Tirole (2005) show that a platform can implement negative pricing for one market side, yet retain overall profitability by "internalizing the externalities" (Evans 2003). However, there are cases when the pricing has been set zero and negative, and the platforms still fail to succeed (Salminen 2014). Currently, the economics-based theory of platforms is unable to provide satisfactory explanations, and comprehensive solutions, for POs struggling with the social dynamics that drive platform growth and success. Hence, other explanations are needed to try and understand platforms from alternative perspectives. According to our argument, presented in the following sections, social contract theory provides a sound basis for a framework that accomplishes just this: an insightful look into social dynamics taking place within a platform.

Platform as a Social Contract

Adapting the Concept of Social Contract to Online Platforms

A first, simple definition of a social contract is that it is an agreement between those who govern (governments, rulers, even POs) and those who are governed (subjects, the people), of implicit or explicit nature (Hobbes 2010; Locke 1988; Rousseau 2003). Enlightenment philosophers, such as Rousseau, Voltaire, and other classical thinkers, have further defined and refined the notion of social contract.[3] Historical trials, such as the French Revolution and United States Declaration of Independence, although not solely motivated by the concept of social contract, have shown the value of their thoughts. While we acknowledge that the political and philosophical theories from which the concept of social contract stems have much deeper and

[3] For clarity, we point out that Enlightenment philosophers were obviously influenced by earlier thinkers, dating back to Ancient Greek (e.g., Plato's teachings on the role of the state as a governing body) and Ancient Rome (e.g., the notion of contract which was central in the Roman society).

more ancient roots, dating back to Ancient Greece and Rome and possibly earlier (Johnstone 2011), we focus on the aspect of this notion that designates the guiding principles of the fundamental, binding agreement – in explicit or implicit form – struck between governed and governing people. In this instance, we consider platform as governing structures and platform stakeholders as governed by this structure's purpose and actions. Much like in ancient kingdoms, which formed the context for the early philosophical discussion on social contracts, platforms involve rulers (PO) and subjects. The "subjects" here refer to the platform's immediate stakeholders, including end users paying for the access or use of the platform (demand side) and service providers offering their services, such as Uber drivers or Airbnb hosts (supply side). The stakeholders give away some freedoms in exchange for benefits, such as safe interaction within the platform. The PO, in exchange, pledges to govern fairly in this implicit social contract arrangement.

From here, the parallel we can draw to platforms is to study them as structures engineered to benefit the stakeholders involved in activities taking place in the platform. A platform succeeds as a structure when it leverages and steers platform stakeholders' behavior in order to benefit the structured ecosystem. In some instance, this pattern is characterized as a network effect (Katz and Shapiro 1994). In any case, the governing structure, the PO, provides guidelines to enable governed "subjects," platform stakeholders, and behavior, in order to simultaneously benefit the ecosystem comprising the PO, stakeholders, and other possible complementors. As such, this requires a long-term trust agreement where each party upholds its promise to contribute to reaching adequate risk/reward ratios for all stakeholders of the platform, including the PO, while not hindering the counterparts' actions.

Overview of the PaaSC Framework

In the following, we build on social contract literature to conceptualize platform as a social contract (PaaSC). We synthesize the social contract literature to the framework that can be employed to explain why end users and other stakeholders engage in the platform play. These pillars are (1) alignment of stakeholder interests, (2) stakeholder support to the contract existence, (3) economic and social justice, and (4) transparency on expectations. Alignment of stakeholder interests builds on Rousseau's general will of the people to guarantee that an actor has an incentive to adhere to the platform rule rather than acting outside the platform. Stakeholder support to the contract existence refers to the approval by the stakeholders. Economic and social justice, referring to Rousseau's concept of social contract, means the stakeholders are backed up and offered guarantees that defend them against contract violators. Finally, acknowledging the importance of understanding each other's

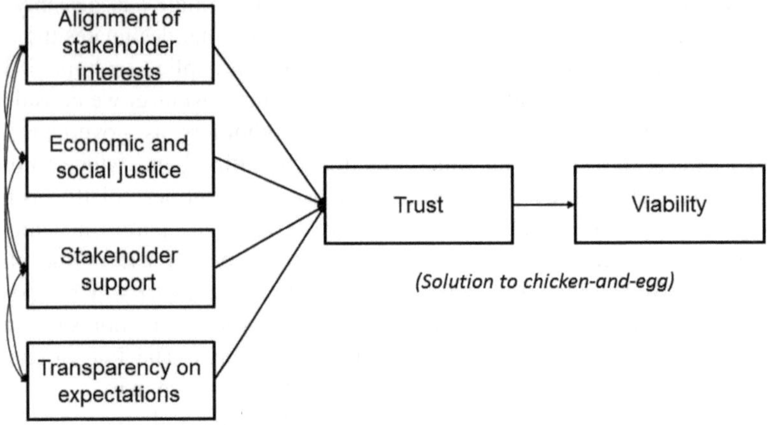

Fig. 1 Conceptual model of the implications of PaaSC framework

expectations is essential for social contracts. Therefore, it is necessary to make the expectations of platform stakeholders transparent in terms of expected effort/cost and benefits of participating in the platform play. Figure 1 shows a visualization of the PaaSC framework.

Satisfying the conditions following from the above four pillars, POs are able to establish trust which is seen as antecedent to viability of online platforms (Zeng et al. 2013). "Viability," in our sense of the word, refers to a sustainable solution to the chicken-and-egg problem. As explained by Salminen (2014), even after the initial solution, the chicken-and-egg problem simply does not "go away" but remains a pressing concern throughout the life cycle of a platform; i.e., at any time individual users can abandon the platform, and, due to the threat of mass escape (cf. negative tipping point by Katz and Shapiro 1985), similar to what took place when users abandoned MySpace, the then dominant platform, the governance of a platform, requires attention to multiple strategic and social considerations (Gawer 2011). Our logic here is that adopting the social contract perspective in the platform governance results in delicate and considerate actions from the PO, thereby decreasing the risk of losing the stakeholders' trust, while resulting in a vibrant, viable platform. It is such actions that can be enhanced by the understanding of the implicit social contract. Note, however, that by viability we refer to the stakeholder interactions in platforms, not financial viability which also has been identified as a critical problem in the platform business model (see Salminen 2014). In brief, portraying platforms as a social contract enables a better understanding of trust in platform markets (which, in turn, is understood as an antecedent to a continuous solution of the chicken-and-egg problem), how each stakeholder's promise toward one another benefits each of them, the interdependency of actors within the platform, and the platform as a social structure.

Four Pillars of PaaSC

Alignment of Stakeholder Interests

First, reflecting Rousseau's "general will" of the people (users), we can argue the following. Although platforms can generally be regarded as dictatorships by governance, because they make the decisions without asking the users, people may rebel against changes and switch the platform (freedom of choice), so ultimately this threat of disconnecting users keeps abuse of power in line. Thus, the users do have a vote of sorts, indirectly in the decisions made by the PO. In Rousseau's view, when the state exceeds its legitimacy, the people will rebel against it and start a new form of government. Pettit (1997) goes as far as to say any social contract is legitimate unless rebelled against. However, the aforementioned low switching cost in the online environment yields platforms sensitive for flight. Rather than high tolerance due to lack of alternatives, a more dominant "glue" for keeping online communities and platforms intact is the existence of network effects (Katz and Shapiro 1985) which make it costly to switch, especially if one has made investments in relationships or content creation that are cumbersome to replace in another platform. While it may be much easier to leave from a platform than from a country, there are hidden chains such as established relationships and other forms of sunk costs that hinder switching behavior.

Reflecting on Hobbes' idea of the state as an impartial judge, we can see it corresponding to the core idea of platforms which is to enable to "build on top of" (Gawer 2011) or enable interaction between its users. The PO remains a neutral party while individual agents negotiate, find each other, and form matches. However, the "natural state" problem of Hobbes emerges if there are no credible threats which would prevent rogue behavior. Therefore, although not directly intervening to interaction between members, the PO needs to provide a trust framework where the members can safely carry out transactions. In practice, this means providing trust features, such as (1) payment mechanisms (e.g., escrow, gateways), (2) evaluation mechanisms (ratings, reviews), (3) penalties (bans, liability, spam detection), and (4) insurance mechanisms (e.g., Airbnb's host insurance). Without the ability to exert power, the platform risks transforming into a chaotic environment of rogue agents.

Let us further illustrate the issue of stakeholder interests. Platforms are built in such a way that the presence or absence of various categories of stakeholders from the platform directly influences the marginal utility of a stakeholder's behavior on the platform. The diversity of demand and supply are bridged by the platform. The services delivered by the platform cannot be perceived as being part of a predetermined set of value chain, in which finality can be anticipated. All things being equal, few hurdles stand in the way of Uber driver who would decide to create his/her own food delivery service, competing with Uber Eat, for example. He would have constituted the first assets of his company through Uber and would cannibalize part of Uber's revenue. Aligning stakeholders' interests, then, is about guaranteeing that they have the incentive of adhering to the platform rule rather than leveraging the platform benefits or functionalities outside of the platform. In our example, Uber would do it by imposing various levels of control on drivers' behavior and capabilities: car models, attitude when welcoming a customer, and derived services at the expense of the driver which the customer would come to expect as part of the driver's value proposition. All of this constitute switching costs which are balancing the betterment of the driver's best interest and that of the platform while he pursues his/her activity.

Stakeholder Support to the Contract Existence

Support takes place when stakeholders take upon themselves to enforce the constraints and preserve the rights given by the social contract of the platform. Stakeholder support is a consequence of the fact stakeholders have digested the rules of the platform and therefore are able to enforce them efficiently. Stakeholder support to contract existence raises several critical objections relating to the platform's value proposition. As long as a stakeholder believes that platform is enforcing the social contract to his/her own benefit, the value proposition encourages him to push the rules of the platform as his/her own. A relevant example to this fact is moderation on any media broadcasting platform. The primary goal of the platform is to transfer as fast and as efficiently as possible the content moderation and curation activities to the content consumer or the content producer side of the platform. For example, YouTube embodies this logic in the "YouTube Heroes" program. The live-streaming video platform Twitch has a similar system whereby moderation can occur within but also across communities, and actually respects both the explicit moderation rules given by the law and by the platform, and implicit rules given by the content producer to his/her own community (which of course do not take precedence on the former).

The issue of moderation is in line with Proudhon (1969), whose view of social contract was based on the idea that individuals retain their personal sovereignty. Applying his view, the social contract does not take place between stakeholders and the PO, but instead between the individuals that refrain from mutual control by giving it to a third party. As shown above, peer moderation is an example of stakeholders internalizing platform rules, and promoting them as their own. However, in the other extreme, where stakeholder support is lacking or negative, there is a risk for revolt. Again, we can draw insight from social contract, recalling the French Revolution. The starting state is a feudal structure that relies on accepting what would be today considered as unfair terms, under rarity of resources. The feudal structure breaks when the scarce resources became more common with the apparition of mercantilism and when the paradigm around protection/security shifted. The initial social contract is broken because the paradigm changes, the stakeholders do not support the contract any longer as it goes against their best interest, hence a revolution to shape a new social contract. Although this is agreeably a broad-stroke painting of this historical event, it lets us draw the following parallel: while the existence of a social contract can be strong enough for the stakeholders to prefer platforms over other contractual agreements (e.g., Uber over a stable employment), once established, the stakeholders expect the PO to respect the social contract and treat them fairly; otherwise, a revolution may take place. A recent phenomenon that could be understood through this logic is the emergence of platform co-ops as an alternative for platforms led by the PO (Scholz 2016). In addition, the press has reported discontent among Uber drivers in several countries. Because the legitimacy of a social contract is contingent upon voluntarism, while any form of coercion delegitimizes it (cf., Spooner 1973), the PO can face hard times in reversing the dynamics of discontent.

Economic and Social Justice

The third pillar, economic and social justice, builds on Locke's idea of the central authority defending the individual against those who seek to do them harm, so that the people can live under the protection of their basic rights. We can easily see the parallels of this idea in the various trust mechanisms employed by modern platforms, backing up and offering guarantees (e.g., Airbnb's apartment insurance) that defend platform stakeholders against violators. Applying Rousseau's thesis, man must "be forced to be free" also in platform context. This justifies the need for certain actions to ensure economic and social justice. It is important to have moderation of spam and quality control in social platforms, where the vocal minority with toxic intents may destroy the quality of discussion and prevent a goodwilling majority to participate, thus threatening the viability of the platform. Unlike government, a platform is not sovereign but in fact a governance system within a broader institutional framework (society). Consequently, information platforms must yield to local laws regulating the acceptable content, for instance, YouTube removing copyright-violating content, Google censoring local results in China, Facebook removing an iconic photo of Vietnam War, and Twitter banning the conservative personage Milo Yiannopoulos.

For example, Reddit has recently censored one of its subgroups, namely, the conservative group, "The_Donald," through introducing algorithmic rules for content visibility. When such control is applied selectively, the fairness of the rule system may erode, and even members of the majority may reject it albeit disagreeing with the minority group. There is a thin red line between censorship and healthy control, most easily identified with discrimination of groups. In the aforementioned case of Reddit, however, one could argue that the discrimination is justified if and only if the group is consistently breaking the rules (laws) of the platform, thereby forming a type of tyranny of minority. Such an effect, in particular, is noted due to Reddit's mechanism of assigning more visibility of its "/all" subdirectory based on votes, which enables coordinated efforts by groups to push forward content of their liking, even to the majority who disagrees with it. Yet, even in this thinking, we may question whether it is justifiable to change the tyranny of minority to the tyranny of majority. Many open questions remain for further analysis of issues of this nature.

A key component of social contract is the ability it gives to platform to arbitrate between various risk/reward ratios proposed to various categories of stakeholders, and how this resonates with the platform's business model. Participation in each platform can be considered as a risk/reward decision. As a result, the platform makes the following promise to platform stakeholders: the investments that stakeholders make in order to abide by the rules will at some point be rewarded by future profits (e.g., through revenue sharing), and the risk/reward associated with belonging to the platform will be adapted to each stakeholder's expectations. Moreover, regardless of the scale of the investment that a stakeholder makes into the platform, this risk/reward ratio will be maintained for him. The violation may originate from other stakeholders or the platform itself, and the risk/reward ratios may remain relatively implicit to an individual. But implicitness exists also at the PO's end. In fact, the level of implicitness may be even a greater problem in the platform setting than in other political contexts, because the power exerted by platforms is hidden behind the algorithm, so that even the PO may be unaware of the exact decision-making taken by its algorithm, when applying complex neural network models (Bose and Liang 1996). At the very granular level of decision-making, therefore, platforms are amoral and cannot be morally judged. Readily available examples of morally adverse outcomes include social injustice and discrimination within Google, Airbnb, and Uber (Bolukbasi et al. 2016; Dillahunt et al. 2016; Malin and Chandler 2016).

Transparency on Expectations

A critical success factor for platforms is their ability to maintain an adequate level of control over stakeholder involvement in the platform play, in direct relation with the quality of services delivered through the platform. As a result, the fourth pillar of a PaaSC framework is about making explicit the expectations for each sides of the platform play in that matter. The POs propose both an operational and a business model which are made available to stakeholders. In return, they define their expectations for a certain degree of involvement in the platform play. Platform stakeholders, in turn, then expect in exchange a certain reward related to the risk (investment) that this involvement represents.

In social contract theory, the rules (expectations for proper behavior) set by the government are translated into laws, which are the guiding principles of social contract. Laws are necessary as people and government, and as a parallel the various sides of a platform play, need to know what they can expect from each other in order to let their relationship be structured through the social contract. In other words, people need to understand the rules (general will of expected behavior) before they support a system. In today's democracies people understand that they are expected to pay taxes, and the representatives know that they are expected to use the tax revenue in a manner that benefits the taxpayers directly or indirectly. Social contract is in force as long as these expectations are explicit and met. When issues arise, the meaningfulness of social contract is questioned, leading to political apathy ("Politikverdrossenheit" in Germany and "ras-le-bol politique" in France) with an increased disrespect toward governments and lack of support to the system.

Issues also emerge when implicit contract is so strong that the platform stakeholders do not bother to read the TOS. Instead, they rely on the benevolence of the PO. That represents the risk that the TOS and the implicit assumptions of the social contract by the stakeholders may be in conflict. Similarly, as in the lack of better information, the POs may misunderstand the expectations laid by the users. Therefore, the solution for an effective social contract is to make the expectations explicit and transparent. The effort goes beyond the definition of TOS, as it needs to echo in the everyday usage of the platform by the stakeholders. In the everyday usage, expectations are not only directed from the government (PO) to citizens (platform stakeholders) or from citizens to the government but from citizens to citizens. In this respect, platform can be seen be more like a modern community than a traditional nation-state. The community has its own expectations for proper behavior, and the members of the community make these expectations transparent in the everyday usage of the platform.

Case Example: YouTube as a Social Contract

Overview of YouTube

YouTube is a media publishing and monetization platform where content producers benefit from a large freedom of publication and content consumers from a variety of content ranging from rebroadcast of evening news to video game walkthroughs. According to criteria defined by Salminen (2014), YouTube may be defined as a two-sided content platform (consumers-producers) or as a two-sided advertising platform (advertisers-producers). The underlying media monetization is advertisement-based, under a variety of formats primarily consisting of advertisement placed in the videos. In the recent years, YouTube has faced major challenges pertaining to its business model: content diversification, war against copyright infringement, community management, and content curation. While content produces have diversified their skills and breadth of coverage to more professional-looking videos spanning topics such as news, music, movies, comedy, video games, and sometimes any combination of the above and more, the users of the community have proved to be one of the most active consumer communities on the Internet. YouTube's business model hence relies, on one hand, on maintaining the vibrancy of content producers while avoiding legal disputes with large franchise or copyright owners such as Walt Disney, Nintendo, Universal Music, and the likes and on the other hand satisfying the expectations of a consumer base which seldom distinguishes YouTube's errors from those of content producers.

Analyzing YouTube's Business Model Challenges Through PaaSC Framework

Alignment of Stakeholder Interests: High

Initially, YouTube, as many media platforms, bet on freedom of speech to federate its community. It was and to a very large extent still is the platform where anyone can publish almost anything not legally reprehensible and reactions can thrive. Recently however it started to act more directly over the externalities at play within its business model. For instance, YouTube has been promoting since a couple of years the champions among the content producers (most prolific, most viewed, etc.) and addressing directly with them the idea that one can make a living out of publishing content on YouTube. They pushed for better formatted content, increasing the average revenue per minute of content produced for the content producers – and thus that of the platform. YouTube also emphasized user ratings by giving to this measurement an increased importance in search results qualification. Thus, this catered for the emergence of content producer champions by content category (such as the famous YouTuber PewDiePie), the quality or notoriety of their content acting as a gravity well inside their respective category or combination of categories – which we might as well refer to as their own value proposition. While this has somewhat come to the expense of content discovery and the emergence of new content producers, it gave the platform more credibility and installed an adequate basis for an ecosystem life cycle where each stakeholder could satisfy its own interest: consuming content appropriate to its taste and quality expectations, producing content that would be pushed through a somewhat fair system allowing the content to meet its audience, and for content producers and content consumers alike, the ability to interact with each other, etc.

Stakeholder Support to the Contract Existence: Somewhat Good

YouTube remains the leading video publishing and monetization platform and as such remains very attractive for any content producer willing to distribute its content to a wide audience and build a sustainable revenue stream from its activity. On the content consumer side, YouTube still represents the best aggregation of media content available on the market, and it is worth noting that both categories of users remain very implicated in testing and assessing the relevance of the choices of YouTube. Even though the attempts of YouTube at implicating them directly into the inner workings of its business model, such as the YouTube Heroes initiative, are still nascent, their support to establishing a robust and viable trust framework for the platform is undeniable and at this point remains one of YouTube's biggest strengths. YouTube has made attempts at producing derived services from its core assets – which obviously are the content search and matching functionalities associated to the media platform. For instance, the short-lived attempt at setting a streaming service potentially competing with Twitch, now a service called YouTube Gaming with lingering activity, seems to have demonstrated that YouTube's brand strength does not bear enough weight to appeal to the content consumer.

Economic and Social Justice: Low

Traditionally, YouTube has given an opportunity for anyone to become famous. The platform has allowed content generators to become professional YouTubers. However, when one takes a look at the increased backlash following YouTube's recent decisions, and in particular the vast unpopularity of the YouTube Heroes initiative, where YouTube is perceived by its community as siding always more with the media giants and large copyright owners rather than the smaller creators abounding on the platform, and giving way to increased censorship on the platform, one can reasonably ask itself whether this strength will last or will in the midterm cease to be one of YouTube's differentiators. Again, it comes down to algorithms and the hectic topic of video demonetization. If the algorithms are set to favor content from certain YouTube celebrities, well-paying sponsors, and established media brands, YouTube is in danger to lose its grip on economic and social justice. When a video is flagged as infringing a copyright owned by a third party, either one of two things can happen: the video gets demonetized or the revenue gets reallocated to the rightful owner of the copyrighted material. Either way, the revenue stream is cut for the content creator. While this seems like a fair situation, it does not consider, for instance, to what extent does the "entertainment" dimension added on top of an existing copyrighted content constitute a fair use of said content. Overall, there needs to be more actions and structures that back up and offer guarantees for content creators that defend them against social contract violators.

Transparency on Expectations: Low

YouTube maintains a trust framework between content creators and content consumers on one hand, and between YouTube (the platform) and content creators on the other hand. The trust relationship between the platform and the content consumer is not emphasized, and one of the major challenges YouTube is facing is that the real heralds of its business model toward content consumers are the content creators. While YouTube has made sizable efforts regarding the transparency on its terms and conditions, on its definition of "fair use" of copyrighted content, or even on explaining the risk and rewards associated to maintaining a sustainable activity as a content producer, a real opacity remains as it leverages proprietary algorithms. More than the content of these algorithms, used for practices such as content ID which are core to YouTube's business model, it is the method through which they are tested and improved that add yet another level of complexity and obscures their nature. For instance, a certain level of automation seems to be in the way algorithms improve their qualification of what is and what is not copyright infringement.

Despite the aforementioned challenges, it is undeniable that YouTube's potential remains tremendous, specifically because in spite of the feud between the platform and the content creators, the ecosystem is more professionalized and vibrant than ever. Amateurs are learning from their experience, increasing their media production skills and professionalizing their use of the platform. In addition, established media brands are leveraging the platform as a broadcasting channel.

Conclusion and Discussion

Overview and Theoretical Contribution

In this research, we introduced a conceptual framework for the analysis of online platforms through the lens of social contract, namely, platforms as a social contract (PaaSC). We built on the works of Enlightenment philosophers to define four pillars of PaaSC that explain trust between platform and its stakeholders. We also demonstrated how this PaaSC framework can be employed to analyze a platform business case by both practitioners and scholars interested in in-depth analysis of platforms. As a consequence, we contribute to the platform literature by deviating from the currently dominant techno-economic perspective to the socio-philosophical view that we see capable in providing answer to the acute problems of POs, such as maintaining a sustainable solution to the chicken-and-egg problem which has been shown to have a persistent nature, taking new shapes beyond user's registration to the online platform (Salminen 2014) and is of high strategic importance to online platforms across all verticals (Rochet and Tirole 2003; Caillaud and Jullien 2003).

In particular, the PaaSC framework introduces novel conceptual angles to analysis of platforms, including (1) alignment of stakeholder interests, (2) stakeholder support to the implicit contract, (3) economic and social justice, and (4) transparency of expectations. Through these concepts, the multidimensional nature of active use of a platform can be better understood than by simply assuming economic rationality or utility derived from the existence of network effects – which, as shown by Salminen (2014), remain theoretical for many unsuccessful online platform startups. In addition, understanding these dimensions results in the logical conclusion of devising practical strategies and actions that correspond with the "real" social dynamics that drive the success and growth of a platform, from an "enlightened" perspective. For example, prior theory on platforms excludes from its analysis the existence of social justice, i.e., perceived fairness of a PO's actions by the stakeholders. Yet, such a social dimension evidently influences based on what is generally known of human behavior: if the platform is considered by the users to operate unfairly, it may reach its demise regardless of efficient pricing strategies and a large extant user base (network effects). If using merely economic explanations, POs and scholars may be left puzzled when this kind of seemingly unexpected events take place – yet, recent history, for example, the case of MySpace, has shown that no platform is protected by the mere existence of network effects. To capture the multitude of factors driving the growth and success of platforms, not due to its limited scope (Shy 2011; Salminen 2014), the PaaSC framework adds a novel approach in understanding modern online platforms. In brief, the contribution of our framework is that it enables the orchestration of all trust-related stakes, which, above and beyond the immediate analysis proposed in the PaaSC framework, also indirectly impact a platform's revenue and margin, in a single framework which integrates with the platform's governance and overall business model in Osterwalder et al.'s (1998) sense of the concept (i.e., strategic decision making for revenue generation, partner choices, key resources, and other means for execution).

Answers to Research Questions

We posed three research questions dealing with the constituents of online platforms as social contracts, description of rights given away by the users, and the stability of good will as an antecedent of mutual trust between stakeholders and POs. The answers to our research questions are summarized as follows. Regarding RQ1, we have identified constituents of the social contract in the context of online platforms. These four pillars of PaaSC include alignment of stakeholder interests, support to the contract existence, economic and social justice, as well as transparency on expectations, which we see as essential constituents of social contracts for online platforms.

As a response to RQ2, we observe that users partially give away some liberties, such as freedom of speech due to moderation (and in exchange receive safe environment) and privacy (in exchange for free access and use of platforms). There are several reasons why the POs are interested in these rights, depending on the business model applied; at the general level, the PO should identify the rights given out by their users and understand the meaning and importance of those rights to the users, so that the PO is able to treat them with according care. In a similar vein, there needs to be understanding of basic rights to guarantee for all users, including safety, freedom from discrimination and harassment, freedom of expression, and other conventional rights that the users implicitly derive from the larger institutional contract. On top of that, there are also internal, acquired, non-explicit rules that guide PO and platform stakeholders' behavior. Both sides have expectations in the matter toward one another, which are described and can be analyzed through the framework proposed above. These expectations differ from the usual relationship a company maintains with consumers, as both PO and platform stakeholders are operationally involved in the platform's business model, hence making the issue of trust and social contract analysis all the more critical for the platform play at every level: strategy, brand, operations, technology, etc. Naturally, we also herein take into account rights and obligations which relate to ground rules or foundations for interaction – those being more common in consumer businesses, such as safety, protection of property (data), and guaranteeing safe transactions and interactions between the users, so that trust can be established, both in the direction of the platform as the institution and toward other users.

Regarding RQ3, we note that the trust in the benevolence of the PO is a logical consequence of the voluntary relationship between it and the stakeholders. In other words, the rule of the PO is accepted as a legitimate control only as long as it protects the basic rights of the stakeholders (which are not defined by the TOS, but by the implicit social contract), while not going too far in its exploitation of the rights given by the users (e.g., some loss of privacy). Therefore, we contend that walking this line constitutes a fine balance; trust in PO is described as a continuum rather than a binary choice, as the PO wants some rights (e.g., private information for ad targeting) which the users will provide, *up to a point*. Determining that point, as said, requires careful maneuvering. Moreover, the protection of the basic rights may not be enough for getting the users to barter away their other rights, because the users also require an additional benefit for their use of the platform. That is, it follows from the logic of embeddedness that, given the surrounding rights are satisfied, whatever basic rights there are in the platform, they are secondary to the surrounding rights. In conclusion, the implications for the PO are threefold: (1) protect the basic rights, (2) do not overexploit the rights given in exchange, and (3) provide additional benefit. Satisfying these constraints, the PO may rest assured the social contract will be kept both sides.

Managerial Implications

Our inquiry is especially relevant for situations where the PO has seemingly solved the chicken-and-egg problem, i.e., there is critical mass of users to allow liquidity, but is now facing the challenge of sustaining the user satisfaction and engagement in the long run. As noted by Salminen (2014), the chicken-and-egg problem is of dynamic nature, meaning that even after initially solving it, the PO must maintain the level of activity taking place in the platform. This feature requires its solutions to be dynamic as well, i.e., creating adaptive and creative approaches to maintain fairness and trust perceived by platform stakeholders. The continuous, day-to-day engagement of the stakeholders is a requisite for accomplishing the business goals of a platform. However, some parts of the literature and also POs in practice may ignore that condition, resulting in short-sighted decisions. At times, there is too high reliance in network effects magically keeping the stakeholders engaged. Envisioning the platform as a social contract, where the parties have benefits in exchange for some rights, may help POs strategically plan mechanisms that are compatible with the social contract. This is particularly important because such a contract tends to be implicit and delegate (e.g., users may not be vocal, but revolt and leave the platform). In the cases where users are vocal, it makes sense to listen to them and understand where their concerns originate. For example, changing the user interface may constitute a violation of the contract from the user's perspective, but the PO who is unaware of the nature of this implicit contract is clueless in the face of rants and flight from the platform. There is a risk of becoming a dictator instead of being Voltaire's benevolent dictator.

According to Locke, individuals from a state provide a "neutral judge" that acts in protection of lives, freedoms, and property rights of those who live in it. But the notion of neutrality requires critical examination, as there is room for contradiction. For example, should the PO remain neutral and enhance freedom, while simultaneously allowing the spread of disinformation or "fake news," or should it seek to protect its users from such disinformation? This "neutrality dilemma" is central in the treatise of platforms as social contracts. The degree of control is an important question for maintaining a healthy culture of participation. Too loose application of rules results in lack of etiquette and degrades content sharing and creation of quality content, whereas too strict applications awake questions of censorship and authoritarianism. Finally, we believe the PaaSC framework can be further developed as a managerial tool to improve the governance of platforms. By applying it in their private analyses, POs can strategize for internal threats but also for external threats such as platform cooperatives.

Policy Implications

Regarding policy implications of our work, as we move toward the "Age of Platforms," the concept of property rights is criticized as its importance is somehow faded out by that of what can be called "control right." In a recent report of the French "National Council,"[4] the lawyer and international expert on property rights Alain Bensoussan claimed that property rights in a digital age, and when it comes to platforms, have to be revised. He claimed that it has basically no legal ground upon which to be based when it comes to personal data, for instance. Instead, he proposed to design a "digital dignity" right which would serve as some kind of constitutional principle to design a proper property right in a virtual world, that would not be enforced by companies (as it today via platforms, such as Facebook or Twitter) but by actual legislative bodies. The future of platforms as societies embedded in societies is an open question, as well as their relationship to the wider institutional framework. Scholars and other stakeholders interested in platforms are gearing up for interesting times.

[4] "Conseil d'Etat," a nonexecutive consultative body providing guidance or advice to the state in economic and social matters among other topics

Future Research Directions

As postulated in this research, the PaaSC framework provides a practical and theoretically sound framework of analysis for social phenomena taking place in platforms. However, the development and validation of the framework can benefit from further research, e.g., inspecting the awareness of POs and stakeholders of the four pillars of the PaaSC and showing its value in real decision-making situations faced by POs. Through an empirical research, we could also see to what extent the four pillars of PaaSC define the amount of trust between the platform and its stakeholders, e.g., by defining a scale to measure its dimensions through a survey-based study. This study also encourages scholars to study the more concrete legal arrangements and economic aspects and discuss what role they play in strengthening the four pillars of PaaSC. This would further validate the use of social contract as an abstraction dealing with ways of how to organize the platform against more simple techno-economic transactional models. We encourage future studies to analyze the effect of structurally different compositions of social contracts on platforms' viability and performance. We acknowledge that there may be cultural and geographic variation in how POs and platform stakeholders formulate and interpret social contracts vis-à-vis formal contracts. While analyzing such variables was beyond the scope of this research, cultural contextualization provides an interesting avenue for future research on platforms.

References

Anderson, S. P., & Gabszewicz, J. J. (2006). The media and advertising: A tale of two-sided markets. In *Handbook of the economics of art and culture* (Vol. 1, pp. 567–614). Philadelphia: Elsevier.

Armstrong, M. (2006). Competition in two-sided markets. *RAND Journal of Economics, 37*(3), 668–691.

Birke, D. (2008). *The economics of networks – A survey of the empirical literature* (SSRN scholarly paper, No. ID 1264662). Rochester: Social Science Research Network.

Bolukbasi, T., Chang, K.-W., Zou, J. Y., Saligrama, V., & Kalai, A. T. (2016). Man is to computer programmer as woman is to homemaker? Debiasing word embeddings. In *Advances in neural information processing systems* (pp. 4349–4357).

Bose, N. K., & Liang, P. (1996). *Neural network fundamentals with graphs, algorithms and applications*. New York: McGraw-Hill.

Caillaud, B., & Jullien, B. (2003). Chicken & egg: Competition among intermediation service providers. *RAND Journal of Economics, 34*, 309–328.

Casey, T. R., & Töyli, J. (2012). Dynamics of two-sided platform success and failure: An analysis of public wireless local area access. *Technovation, 32*(12), 703–716.

Coase, R. H. (1937). The nature of the firm. *Economica, 4*(16), 386–405.

Colomer, J. M. (2002). *Political institutions in Europe*. London: Psychology Press.

Dillahunt, T., Lampinen, A., O'Neill, J., Terveen, L., & Kendrick, C. (2016). Does the sharing economy do any good? In *Proceedings of the 19th ACM conference on computer supported cooperative work and social computing companion* (pp. 197–200). New York: ACM.

Economides, N., & Katsamakas, E. (2006). Two-sided competition of proprietary vs. open source technology platforms and the implications for the software industry. *Management Science, 52*(7), 1057–1071.
Eisenmann, T., Parker, G., & Van Alstyne, M. (2011). Platform envelopment. *Strategic Management Journal, 32*(12), 1270–1285.
Evans, D. S. (2003). Some empirical aspects of multi-sided platform industries. *Review of Network Economics, 2*(3), 191–209.
Evans, D. S. (2011). How catalysts ignite: The economics of platform-based start-ups. In A. Gawer (Ed.), *Platforms, markets and innovation*. Cheltenham: Edward Elgar Publishing.
Gallopín, G. C., Funtowicz, S., O'Connor, M., & Ravetz, J. (2001). Science for the twenty-first century: From social contract to the scientific core. *International Social Science Journal, 53*(168), 219–229.
Gawer, A. (2011). *Platforms, markets and innovation*. Cheltenham: Edward Elgar.
Hagiu, A. (2006). Pricing and commitment by two-sided platforms. *The Rand Journal of Economics, 37*(3), 720–737.
Hagiu, A., & Wright, J. (2014). Marketplace or reseller? *Management Science, 61*(1), 184–203.
Hagiu, A., & Wright, J. (2015). Multi-sided platforms. *International Journal of Industrial Organization, 43*, 162–174.
Hobbes, T. (2010). *Leviathan: Or the matter, forme, & power of a common-wealth ecclesiasticall and civill* (I. Shapiro, Ed.) (Reprint ed.). New Haven: Yale University Press.
Johnstone, S. (2011). *A history of trust in ancient Greece*. Chicago/London: University of Chicago Press.
Katz, M. L., & Shapiro, C. (1985). Network externalities, competition, and compatibility. *The American Economic Review, 75*(3), 424–440.
Katz, M. L., & Shapiro, C. (1994). Systems competition and network effects. *The Journal of Economic Perspectives, 8*(2), 93–115.
Lampinen, A., & Cheshire, C. (2016). Hosting via Airbnb: Motivations and financial assurances in monetized network hospitality. In *Proceedings of the 2016 CHI conference on human factors in computing systems* (pp. 1669–1680). New York: ACM.
Lampinen, A., Bellotti, V., Monroy-Hernández, A., Cheshire, C., & Samuel, A. (2015). Studying the sharing economy: Perspectives to peer-to-peer exchange. In *Proceedings of the 18th ACM conference companion on computer supported cooperative work & social computing* (pp. 117–121). New York: ACM.
Locke, J. (1988). *Locke: Two treatises of government* (P. Laslett, Ed.) (3rd ed.). Cambridge/New York: Cambridge University Press.
Malin, B. J., & Chandler, C. (2016). Free to work anxiously: Splintering precarity among drivers for Uber and Lyft. *Communication, Culture & Critique, 10*, 382–400.
Mas, I., & Radcliffe, D. (2011). Scaling mobile money. *Journal of Payments Strategy & Systems, 5*(3), 298–315.
Oakeshott, M. (1991). *On human conduct*. Oxford: Clarendon Press.
Osterwalder, A., Pigneur, Y., & Tucci, C. L. (1998). Clarifying business models: Origins, present, and future of the concept. *Communications of the Association for Information Systems, 16*, 1–25.
Parker, G. G., & Van Alstyne, M. W. (2005). Two-sided network effects: A theory of information product design. *Management Science, 51*(10), 1494–1504.
Pettit, P. (1997). *Republicanism: A theory of freedom and government*. Oxford: Oxford University Press.
Proudhon, P.-J. (1969). *Selected writings of Pierre-Joseph Proudhon*. Garden City: Anchor Books.
Rifkin, J. (2014). *The zero marginal cost society: The internet of things, the collaborative commons, and the eclipse of capitalism*. New York: Palgrave Macmillan.
Rochet, J.-C., & Tirole, J. (2003). Platform competition in two-sided markets. *Journal of the European Economic Association, 1*(4), 990–1029.

Rochet, J.-C., & Tirole, J. (2005). *Two-sided markets: A progress report* (IDEI working paper, No. 275). Toulouse: Institut d'Économie Industrielle (IDEI).

Rousseau, J.-J. (2003). *On the social contract* (G. D. H. Cole, Trans.) (Dover Thrift Editions ed.). New York: Dover Publications.

Salminen, J. (2014). *Startup dilemmas – Strategic problems of early-stage platforms on the internet*. Turku: Turku School of Economics.

Salminen, J., & Teixeira, J. (2013). Fool's gold? Developer dilemmas in a closed mobile application market platform. In *Proceedings of the ICEC 2013 the 15th International Conference on Electronic Commerce*, Turku.

Sawhney, M., Verona, G., & Prandelli, E. (2005). Collaborating to create: The internet as a platform for customer engagement in product innovation. *Journal of Interactive Marketing, 19*(4), 4–17.

Scholz, T. (2016). *Platform cooperativism. Challenging the corporate sharing economy*. New York: Rosa Luxemburg Foundation.

Seamans, R., & Zhu, F. (2013). Responses to entry in multi-sided markets: The impact of craigslist on local newspapers. *Management Science, 60*(2), 476–493.

Shy, O. (2011). A short survey of network economics. *Review of Industrial Organization, 38*(2), 119–149.

Smith, A. (1994). *The wealth of nations* (E. Cannan, Ed.) (Sixth Printing ed.). New York: Modern Library.

Spooner, L. (1973). *No treason: The constitution of no authority*. Colorado Springs: Ralph Myles Publisher.

Susarla, A., Barua, A., & Whinston, A. B. (2009). A transaction cost perspective of the "software as a service" business model. *Journal of Management Information Systems, 26*(2), 205–240.

Williamson, O. E. (1981). The economics of organization: The transaction cost approach. *American Journal of Sociology, 87*(3), 548–577.

Zeng, Q., Zhang, Y., & Jing, N. (2013). Research on the mechanism of trust in the open online sourcing marketplace. In J. Zhang, Z. J. Wang, S. R. Zhu, & X. M. Meng (Eds.), *Information technology applications in industry* (Pts 1–4, Vol. 263–266, pp. 2732–2737). Stafa-Zurich: Trans Tech Publications.

Expanding the Platform: Smart Contracts as Boundary Resources

Kristian Lauslahti, Juri Mattila, Taneli Hukkinen, and Timo Seppälä

Abstract Platform businesses are born global, with instant access to global markets. Thanks to the algorithmic, self-executing and self-enforcing computer programmes known as smart contracts, platform businesses now also have instant access to global capital markets from birth. However, the legal status of these smart-contract-enabled funding mechanisms and smart contracts in general is not well defined. In this article, we analyse how well the formation mechanisms of the general principles of Finnish contract law can be applied to the technological framework of smart contracts. We find that depending on the case, smart contracts can create legally binding rights and obligations to their parties. We also observe that contracts have not been formerly perceived as technical boundary resources in the sense that platform ecosystems could foster broader network effects by opening their application contracting interfaces to third parties.

Keywords Platform · Smart contract · Boundary resource · Contract law · ICO

K. Lauslahti
The Research Institute of the Finnish Economy (ETLA), Helsinki, Finland

J. Mattila (✉) · T. Seppälä
The Research Institute of the Finnish Economy (ETLA), Helsinki, Finland

Aalto University School of Science, Espoo, Finland
e-mail: juri.mattila@etla.fi; timo.seppala@etla.fi

T. Hukkinen
Aalto University School of Science, Espoo, Finland

© The Author(s) 2018

A. Smedlund et al. (eds.), *Collaborative Value Co-creation in the Platform Economy*, Translational Systems Sciences 11,
https://doi.org/10.1007/978-981-10-8956-5_4

From Digital Contracts to Lex Cryptographia

In 1994, American cryptographer Nick Szabo published an article in which he outlined the concept of smart contracts.[1] Szabo defined smart contracts as machine-readable transaction protocols which create a contract with predetermined terms.[2] In its simplest form, a smart contract is a machine-readable programme, written in code that will execute itself when a set of predetermined terms are met.[3]

Regardless of the advanced ideas and the advanced concept, the IT infrastructures of the era were considerably behind the level required to bring Szabo's vision to reality, and the time was not yet ripe for practical experimentation.[4] Now years later, the concept of smart contracts has resurfaced as the technological development has caught up with the vision.[5] New technical advances in blockchain technology[6] have enabled the transition from automated digital contracts to truly autonomous smart contracts, capable of self-execution and self-enforcement.

The relationship between platforms,[7] blockchain-based smart contracts and contract law creates an interesting research environment in which the traditional definition of contracts is placed under review as coded programmes begin to administer transactions. Moreover, legal research on blockchain technology has been said to lead to the development of a new legal field which can be described as *Lex Cryptographia*, or crypto law.[8] Determining the legal nature of smart contracts is in

[1] The original text *Smart Contracts* is available at http://szabo.best.vwh.net/smart.contracts.html (17 June 2016). The text *The Idea of Smart Contracts* published in 1997 took the idea of smart contracts further: http://szabo.best.vwh.net/smart_ contracts_idea.html (17 June 2016).

[2] By transaction protocols Szabo meant protocols between different devices, which achieve the so-called Nakamoto consensus. Szabo 1994: "A smart contract is a computerised transaction protocol that executes the terms of a contract". According to a newer definition, a smart contract is "a set of promises, specified in digital form, including protocols within which the parties perform on these promises" Szabo 1996.

[3] It is noteworthy that smart contracts do not need artificial intelligence to work, regardless of what their name may suggest.

[4] See, e.g. Glatz 2014: *What are Smart Contracts? In search of a consensus.*

[5] For example, http://tech.cornell.edu/news/smart-contracts-the-next-big-blockchain-application (23 August 2016)

[6] For this chapter, we define blockchain technology as the cryptographically concatenated data structure and the network architecture described by Nakamoto (2008) which entails a proof-of-work consensus protocol and employs cryptographic tokens of value, more commonly referred to as cryptocurrency.

[7] For this chapter, we define a platform as an IT system that enables a multisided market environment where different market sides can perform value-adding activities that are complementary to one another and which are governed by boundary resources. For further platform literature, see, e.g. Cusumano and Yoffie 1998; Cusumano and Gawer 2002, Cusumano 2005, 2010; Parker and Van Alstyne 2005; Eisenmann et al. 2006; Gawer and Henderson 2007; Boudreau and Hagiu 2008; Gawer and Cusumano 2008; Gawer 2009; Baldwin and Woodard 2009; Tiwana et al. 2010; Yoo et al. 2010; Kenney and Pon 2011; Parker and Van Alstyne 2014; Hagiu 2014; Pon et al. 2014, 2015 Parker et al. 2016.

[8] A new legal field *Lex Cryptographia* focuses on rules which are managed through self-executing smart contracts and decentralised autonomous organisations. See Wright and De Filippi 2015, p. 48.

fact a key theme in the surrounding discussion[9] in which they have been increasingly assessed as legally relevant activity.[10] Thus, it should be noted that smart contracts are not only administered by their programming logic or, in other words, the code they contain; they are also inseparably influenced by the state of the law.[11]

Platform businesses are born global, with instant access to global markets. Through recent developments in smart contracts, platform businesses now also have instant access to global capital markets from birth.[12] However, the legal status of these smart-contract-enabled funding rounds and smart contracts in general is not well defined at this point. The techno-economic point of view has traditionally been selected as the dominant way for understanding technological disruptions and their effects. In recent years, however, legal regulation has also been increasingly understood as an equally important factor in developing innovations in the platform economy.[13] This calls for a systematic review of the legal doctrinal composition of smart contracts within the context of an established legal framework.

Frameworks of the same historic background – such as those with their roots in the Romano-Germanic legal tradition – share more commonalities with each other than with systems descending from another historic background, such as those based on the common law legal tradition. Therefore, there are differences in the compositions of contractual mechanisms in different legal frameworks. Thus, an all-pervasive systematic review cannot be covered in one research article. Instead, the legal doctrinal composition of smart contracts must be evaluated for each legal framework individually.[14]

---

[9] About the nature of smart contracts more generally: "They are defined variously as 'autonomous machines', 'contracts between parties stored on a blockchain' or 'any computation that takes place on a blockchain'". Many debates about the nature of smart contracts are really just contests between competing terminology [...]", http://www.coindesk.com/making-sense-smart-contracts/ (23 August 2016).

[10] Glatz 2014: "It is however undeniable, that smart contracts have to be classified as legally relevant behavior. [...]". See also Koulu 2016, p. 54.

[11] See, e.g. *Blockchain 2.0, smart contracts and challenges*: http://www.twobirds.com/en/news/articles/2016/uk/blockchain-2-0--smart-contracts-and-challenges#1 (23 August 2016).

[12] Global capital markets have been enabled by a phenomenon around blockchain technology generally referred to as *token sales* or *initial coin offerings (ICO)*. For more information on ICOs, see, e.g. Conley 2017.

[13] See Chander 2014.

[14] It is noteworthy that the legislation concerning electronic contracts and information society in general has been harmonised to an extent on the EU level (e.g. in Finland's case, see the Finnish Information Society Code, 917/2014). In the case of characterising the legal status of a single service provider utilising smart contracts, a more systematic review on the EU level could be in order. However, as this chapter focuses on answering a more basic question about the contractual applicability of smart contracts in general, this harmonised legislation does not fall within the scope of this research. Similarly, on the national level, other mandatory provisions (e.g. distance selling and distance selling of financial services) may have legal implications concerning smart contracts. These include inter alia, the conclusion of special consumer contracts and other public-law-oriented provisions in acts such as the Money Collection Act (255/2006), the Crowdfunding Act (734/2016) and the Securities Markets Act (746/2012).

In this chapter, we examine the relationship between blockchain-based smart contracts and Finnish contract law.[15] The main research question herein is *whether or not legal acts can be concluded with smart contracts under Finnish contract law*.[16] In order to provide an answer, it must, first of all, be clarified how the general doctrines of contract law are applicable to these new smart contracts in terms of conferring rights and imposing obligations on parties. Secondly, it must be determined whether all smart contracts constitute contracts in themselves, or whether there are internal requirements for their legal significance.[17]

We conclude the chapter with a discussion on the implications of our findings on multisided platforms and the platform economy at large. Smart contracts are a clear example of how some social boundary resources of platforms are developing in an increasingly technical direction and should be perceived as technical enablers, similarly to technical boundary resources.[18] Contracts in themselves have not been formerly perceived this way, in the sense that the network effects of a platform ecosystem could be boosted by opening up "application contracting interfaces". This would mean, for instance, the application of even further automated digital contracting mechanisms, process automation that reaches further beyond a company's own information systems, as well as further automated and more dynamic networks of contracting parties.

This chapter continues as follows: In the second section of the chapter, we will outline the definition of smart contracts and discuss the creation of a smart contract from the perspective of contract law. In section three, we will seek to answer the

[15] Due to the notable proximity of the Finnish legal framework to those of the other Nordic countries, some analogies thereto most likely are justified.

[16] In this chapter, the research method of choice is mainly legal doctrinal (or legal-dogmatic) research, the main focus being on the research of current positive law – but in our case, examined in a broader context of the platform economy. See Hirvonen 2011, pp. 21–23 and 28–30.

[17] In this publication, it is not possible to discuss central guidelines not related to the content of smart contracts or the interpretation of such content. Questions regarding parties and legal entities in general have also been left undiscussed apart from a few mentions. Furthermore, the question of which country's national legislation should be applied to smart contracts is also interesting. Smart contracts exist in a blockchain that functions in a decentralised environment, and the parties (of which there may be several) may be completely unknown to one another. Therefore, it may not be clear which jurisdictions are relevant to the contract unless specifically referred to in its terms. It is important to study this question, but it is likely that any factual solutions to this issue will only be found through practice. In addition to the questions above, it is also important to consider how programming is viewed by Finnish contract law. Is it possible to equate the programming of a smart contract to a middleman, comparable to counsel drafting a traditional contract? While these interesting questions are mostly brushed aside in this text, it should be noted that the importance and role of programming will be an increasingly important topic in the future.

[18] In platform literature, boundary resources are the operational regulations and technical tools and interfaces governing the interaction between the platform owner and the platform participants. They can be used either to encourage platform development or to restrict it in places where the platform owner wishes to maintain control over the developmental direction of the platform. These resources are sometimes divided into technical and social boundary resources. For further information, see, e.g. Gawer 2009; Yoo et al. 2010; Ghazawneh 2012; Ghazawneh and Henfridsson 2013.

question of whether legal acts can be concluded with smart contracts, and finally, in section four, we will discuss the impact of smart contracts in the context of development trends of digital platforms and the surrounding ecosystems.

The Nature of Smart Contracts

Smart Contracts

A fully established definition for smart contracts has yet to be formed. According to Nick Szabo, creator of the concept behind smart contracts, however, the most primitive example of a smart contract is, in fact, a regular vending machine where transactions are based on simple mechanical automation. The vending machine, due to its physical design, accepts coins, hands over the selected item and finally returns the change. The machine, therefore, completes the transaction on its own when the necessary prerequisites are met – that is, a sufficient amount of money has been deposited into its slot. Anyone in possession of a sufficient amount of coins and with the desire to purchase one of the items for sale is capable of becoming a contracting party in this type of a transaction. Additionally, since the items for sale are situated within the vending machine, it is capable of protecting the logic of its proposed contract from unauthorised changes.[19]

Much in the same way as vending machines, digital smart contracts can essentially be characterised as cryptographic "boxes" containing value that only unlocks upon the fulfilment of the preconditions determined in their design.[20] In other words, smart contracts are automated mechanisms under the control of which assets can be deposited and which then autonomously redistribute those assets according to their internal programming logic.[21] As such, smart contracts enable the execution of transactions to be automatically based on data that was not yet available when the contract itself was concluded.[22]

---

[19] Szabo 1994.

[20] Ethereum White Paper 2013.

[21] From a more technical point of view, smart contracts are autonomous programmes situated in a certain address in the blockchain, which can be rerun infinitely and can also be programmed to contain a wide array of business model logics. Once the events specified in the contract take place and the transaction containing data arrives to the address of the smart contract, the distributed virtual machine of the blockchain executes the programming code of the smart contract in question. Ethereum is one example of this type of a blockchain platform with an integrated virtual machine layer which allows programmes to be run in a fully decentralised fashion and thus can facilitate smart contracts. See, e.g. http://ethdocs.org/en/latest/introduction/what-is-ethereum.html (23 August 2016). BBVA Research – Digital Economy Outlook, October 2015, p. 4 (https://www.bbvaresearch.com/wp-content/uploads/2015/10/Digi-tal_Economy_Outlook_Oct15_Cap1.pdf) (23 August 2016).

[22] Buterin 2014.

Diverging from contracts concluded in the form of action, speech or writing, a smart contract is characteristically a computer programme built in code. Moreover, as currently employed in reality, smart contracts are based on decentralised peer-to-peer networks and reside in a distributed network database known as a blockchain.[23] In order to implement a contractual arrangement as a smart contract in practice, the terms of the proposed contract are formulated in programming language, after which the smart contract is deployed in the blockchain. Once deployed, the distributed blockchain network executes the smart contract automatically without the assistance of the contracting parties whenever the conditions outlined in the code of the smart contract are met.

Due to their decentralised nature, smart contracts are often said to be *self-executing* and *self-enforcing*. In other words, they differ significantly from conventional forms of digital contracts, such as clickwrap contracts, in that they do not require a centralised trusted party to administer the execution of the contract in the digital world.[24] Moreover, blockchain networks are capable of preventing unauthorised changes to the internal logic of the smart contracts in their distributed database. Therefore, no party or authority has the power to prevent such networks from executing the smart contracts in their original form.[25]

Based on all the characterisations above, we define smart contracts for this chapter as digital programmes that

(a) Are written in computer code and formulated using programming languages
(b) Are stored, executed and enforced by a distributed blockchain network
(c) Can receive, store and transfer digital assets of value
(d) Can execute with varying outcomes according to their specified internal logic

From this definition, it is easy to see that the established term for describing such cryptographic boxes of value, namely, "smart contracts", can be quite misleading, as their smartness as well as their contractual nature can both be called into question. In essence, smart contracts are merely automatic programmes built in code and deployed on a blockchain to perform logical processes. Thus, the term "smart contracts" is also commonly used in connection with many other types of programmes situated in the blockchain and not only those resembling a formal agreement.[26] Smart contracts are also capable of actions such as collecting data from outside resources (API oracles) and processing it according to the terms specified in their

[23] At the moment, the most prominent of such platforms for smart contracts is a blockchain known as Ethereum (see https://www.ethereum.org/ (23 August 2016). For additional information on blockchain technology in general, see, e.g. Mattila (2016).

[24] For more information on the role of smart contracts in the evolution of digital contracts in general, see, e.g. (Werbach and Cornell 2017; Kõlvart et al. 2016).

[25] Mattila 2016, p. 15. The irreversibility of some contracts may prove to be a problem in some situations. This issue will, however, not be discussed further in this text.

[26] See, e.g. Stark, Josh: *How Close Are Smart Contracts to Impacting Real-World Law?* http://www.coindesk.com/blockchain-smarts-contracts-real-world-law/ (23 August 2016).

programming logic and executing concrete varying outcomes based on the results of this procedure.[27]

Nonetheless, it is possible to give smart contracts characteristics that can be likened to those of conventional contracts – at least from a theoretical viewpoint – by formulating their internal logic accordingly.[28] In such cases, smart contracts begin to show contract-like characteristics once digital assets have been transferred to their control and once they are transferred again in order to redistribute them according to the prespecified criteria.[29]

Contract Law and the Interpretation of Smart Contracts

Contracts are a key legal instrument for private operators as they execute changes in their legal relations or try to prepare for future turns of events. Contracts also enable organised collaborative activity and are often used to carry out economic activity.[30] The definition of the term "contract" contains a number of different meanings. First of all, the term may refer to the conclusion of the agreement itself, therefore describing the parties' commitment to the contract. Secondly, it may refer to the contents of the agreement, therefore determining the parties' rights and obligations in relation to one another. Thirdly, it may refer to the actual document in which the terms of the contract have been specified.[31]

Contract law is traditionally non-mandatory. In other words the parties can disregard certain rules of presumption by implementing their own terms. This principle of freedom of contract is the premise from which Finnish contract law also sets out. For a number of reasons, however, freedom of contract is restricted by certain mandatory rules regarding the content of agreements.[32] The main principle is, nonetheless, that parties can exercise full freedom in deciding whether to enter into a contract, with whom, in what manner and with what terms. The right to decide on

---

[27] BBVA Research – Digital Economy Outlook October 2015, p. 4 (https://www.bbvaresearch.com/wp-content/uploads/2015/10/Digital_Economy_Outlook_Oct15_Cap1.pdf) (23 August 2016)

[28] Koulu 2016, p. 65: "[…] the smart contract operates with a similar logic to 'traditional' contracts: the will of both parties to enter the agreement is needed in order for it to be valid".

[29] It must be noted, however, that the aforementioned course of events is only a presumption, and the smart contract can also remain at a stage where it functions purely as a re-router built to transfer data or, for instance, the contents of one crypto-wallet to another (Bourque and Fung Ling Tsui 2014, p. 10). The legal status of such smart contracts can indeed be questioned with good reason, at least from the perspective of contract law. Therefore, their interpretation would seem to require case-by-case evaluation.

[30] Hemmo 2003, p. 4; 2006, p. 27.

[31] Saarnilehto et al. 2012, p. 310.

[32] Hemmo 2003, p. 77.

the dissolution of a contract has also been considered an important, yet separate, part of freedom of contract.[33]

In addition to the principle of freedom of contract, the Finnish legal system also acknowledges the principle of *pacta sunt servanda*, that is, agreements must be kept.[34] Various sanction mechanisms also make it necessary to abide by the contracts one has entered into, since the other party has the opportunity to claim damages or enforce the contract by the help of the authorities.[35]

In this publication, we will address contracts as individual agreements concluded between rational and equal private parties with the main purpose of organising economic legal relations. Due to practical reasons, our presentation of Finnish contract law will be limited to a rather general level, focusing on the mechanisms leading to the conclusion of a contract. Our goal in this endeavour is to analyse through doctrinal research[36] and as straightforwardly as possible those aspects of contract law which are relevant to the interpretation of smart contracts. This perspective leaves out several significant legal themes which we are not able to explore in this publication. Since there has been little research on smart contracts, this type of approach is necessary in order to define them and assess them in a legal context.

Legal Acts, Declarations of Intent and Contracts

The relationship between legal acts and contracts has so far been widely discussed in Finnish legal literature, and scholars have tried to find differences in the meanings of these terms. Recently, however, these terms have increasingly often been used as synonyms for each other,[37] although Finnish legislation still includes well-established expressions which utilise the term legal acts. In this publication, we will adhere to the practice of using the two terms synonymously.

Consent, declaration of intent and the purpose that this intent becomes known to the other party have all been considered sine qua non for a legal act. Consent refers to a party's free will to become bound by the contract. In addition, this consent must become known to the recipient in one way or another.[38] Declaration of intent refers

[33] Hemmo 2003, pp. 69, 72 and 75–77.

[34] Section 1(1) of the Finnish Contracts Act (228/1929): "An offer to conclude a contract and the acceptance of such an offer shall bind the offeror and the acceptor as provided for below in this chapter".

[35] Hemmo 2003, p. 14; Saarnilehto 2009, pp. 161–163.

[36] Doctrinal research, or legal dogmatics, attempts to study law as it currently stands. See more: Hirvonen 2011, pp. 21–26.

[37] For example, Mika Hemmo has used these two terms as synonyms. For more, see Hemmo 2003, pp. 10–11 and Hemmo 2006, p. 26.

[38] Saarnilehto et al. 2012, p. 323.

to the expression of a party's[39] free will as a prerequisite to the conclusion of a contract. Both parties are free to decide what their will is and how they are bound to the decision. Although the declaration of intent should by principle be directly addressed to a certain other person or group, even a declaration of intent addressed to a more vaguely specified person or group of people can be seen as valid.[40] This, however, requires a restriction of some sort regarding the targeted group, as entirely unspecified public declarations of intent have by principle been considered non-binding. The reasonable impression that the declaration has had on the recipient has been utilised as a key argument in assessing whether or not the declaration has binding effects. For instance, an advertisement in a newspaper has not as such been considered a sufficient offer.[41] On the other hand, an automat which has been set up with its for sale items and relevant information (regarding prices, methods of payment, products, etc.) may be considered a de facto offer which has been made to a sufficiently limited audience, that is, those in the immediate vicinity of the automat.

The declaration of intent must be expressed clearly. That said, an implied expression of intent is also valid, and intent can be expressed through various forms of communication. The thought or idea of an agreement alone, however, does not constitute a declaration of intent. The method, form and audience of the declaration are not subject to overly strict regulation, and it is in fact sufficient that consent is expressed in one way or another.[42] It is also not imperative to apply an overly strong presumption on the necessity of such a declaration. Not all methods of concluding a contract even require a proper declaration of intent. Additionally, the declaration of intent does not need to be entirely separate from the agreement, as a contract can also be concluded based on passivity or concrete actions.[43] It follows that a party's true will to be bound and some expression of this intent are of key importance.

A contract is a bilateral legal act which establishes rights and obligations for the parties to it. Only the parties to a contract may demand that these obligations should be met. A third party only has this right in certain exceptions.[44] In Finnish jurisprudence, contracts have traditionally been defined as the combination or amalgamation of two or more legal acts requiring one another. In some cases, specific requirements as to form must also be met, or certain actions must be performed before a contract can fully enter into force.[45] The conclusion of a contract is often

---

[39] In Finland, "legal acts" can be concluded by all natural persons (i.e. humans) and legal persons for whom requirements have been set in order to have legal capacity. Questions about legal entities may arise especially in relation to decentralised autonomous organisations but also about the different interpretations relating to the nature of smart contracts. Some researchers have considered smart contracts as agents based on algorithmic contracts acting for and on behalf of their principal or even independent legal entities. See, e.g. Scholz, Lauren Henry: Algorithmic Contracts (draft, 2016) and Bourque and Fung Ling Tsui 2014, pp. 18–19. Questions about legal entities have their own connection to smart contracts, but that will not be considered any further in this text.

[40] Saarnilehto et al. 2012, p. 323.

[41] Hemmo 2006, pp. 78–79.

[42] Saarnilehto et al. 2012, p. 328.

[43] Hemmo 2003, pp. 11–13.

[44] Norros 2007, pp. 1–3

[45] Saarnilehto 2009, p. 3; Saarnilehto et al. 2012, pp. 367–368.

related to the organisation of economic activity.[46] In recent decades, however, the social dimension of contracts has also been emphasised. A reasonable balance in terms of the material content of a contract has been considered a prerequisite for the binding effect of a contract. In addition, parties in a weaker position are not thought to have a very extensive duty to investigate or make enquiries.[47]

Mechanisms for Concluding Contracts

The so-called offer–acceptance mechanism, as it is regulated in the Finnish Contracts Act, is seen as the traditional method for concluding a contract and is based on two legal acts. As contracts are becoming all the more diverse, the offer–acceptance mechanism is not, however, always the most accurate description of the process leading to the conclusion of a contract.[48] Under section 1 of the Contracts Act, the offer to conclude a contract and the acceptance of such an offer are binding in regard to the offeror and the acceptor. The Contracts Act, however, does not apply to contracts of standard form or contracts which require acting upon in order to become effective.[49] The response to the offer must be delivered on time and must accept the original offer as such. The Contracts Act provides that a response that purports to be an acceptance, but includes additions or restrictions, is to be deemed a rejection constituting a new offer directed at the original offeror.[50]

Mechanisms for concluding a contract not regulated by the Contracts Act include contracts concluded through negotiation, implied contracts and tacit agreements. Standard-form contracts are also considered to be formed outside the offer–acceptance mechanism.[51] Aside from contracts concluded via the offer–acceptance mechanism, implied contracts and tacit agreements are the most relevant to smart

[46] This characteristic has at least been heavily emphasised. See Hemmo 2006, p. 24.

[47] See, e.g. Hemmo 2003, pp. 19–24. The so-called social civil justice emphasises the mutual trust between the parties and the principle of equity of contracts. An unreasonable contract or individual term may, therefore, be amended by the court for reasons of equity. This feature of Finnish contract law will most likely be applied to smart contracts as well. Only time will tell, however, whether courts will have the competence to evaluate whether a smart contract written in computer code is equitable.

[48] Hemmo 2003, pp. 96–97.

[49] These kinds of contracts, which require acting upon (the interposition of something), are called real contracts and, in legal literature, have been considered to have very little importance in Finland. "*Reaalisopimuksen sitovuuden edellytyksenä on sopimuksen kohteen luovuttaminen toisen hallintaan*" [For a real contract to be binding the subject matter of the contract must be handed over to the other party's possession]. See Hemmo 2003, pp. 100 and 180–181.

[50] Finnish Contracts Act (228/1929, as amended): http://www.finlex.fi/fi/laki/ajantasa/1929/19290228#L3 (23 August 2016). The Contracts Act includes more detailed provisions about responses given on time, power of attorney and invalidity of juristic acts.

[51] Hemmo 2003, pp. 129–137.

contracts. In addition, smart contracts may contain similar characteristics to contracts requiring acting upon in order to become effective.

Implied contracts refer to a situation where a contract is seen to have been concluded without explicit expressions of intent but rather based on social norms. In these situations a contract has been concluded based on some action, without any oral or written exchanges. Typically these actions have similar qualities to a contract and are part of a prevalent social convention which both parties are deliberately participating in.[52] Examples offered by legal literature of such social conventions could be using public transportation or parking in a paid parking lot. Using an automat has also sometimes been placed in this category. In summary, implied contracts are contracts based on certain facts inducing a contractual relationship but where no explicit offer–acceptance mechanism takes place.

The term "tacit agreements" is also used to describe a slightly similar phenomenon. The term refers to the conclusion of a contract through a situation in which no explicit declaration of intent can be detected, although the parties collaborate in a way that indicates the existence of a contractual relationship.[53] It has been stated in legal literature that it is mostly a matter of taste which term to use.[54,55] When parties collaborate in a way that denotes a contractual relationship, a contract is seen to have been implicitly concluded, even though the method and time of conclusion and the contract itself cannot be shown. Therefore, if parties have commenced action as if the contract were in force, despite the contract's itself remaining in the stage of negotiations or not yet having being concluded, an implicit contract may be in force between the parties. The interpretation of whether a tacit agreement has been concluded is based on overall evaluation, in which circumstances strongly speaking in favour of the existence of a contract can prove that a tacit agreement has entered into force. However, even rather minor arguments against the existence of a contract can relatively quickly lead to the conclusion that no tacit agreement has been reached between the parties.[56] Interpretation should not be too liberal in order to avoid parties being bound to contracts they have not declared their intent for.[57]

According to legal literature, a declaration of intent leading to the conclusion of a contract can be expressed by the parties through the exchange of assets or services with one another. A similar transaction-based interpretation has also been outlined in regard to smart contracts.[58] A declaration of intent by acting upon it can, for instance, take place in the purchase of items from a vending machine. In this case, the proprietor selling items and services via the vending machine has implicitly displayed its desire to conclude a contract with the terms specified by the vending machine. This is supported, for example, by the fact that the proprietor has first had

---

[52] Hemmo 2003, pp. 131–133.

[53] Hemmo 2006, p. 88.

[54] Implied contract or tacit agreement.

[55] Saarnilehto et al. 2012, p. 385.

[56] Hemmo 2003, pp. 133–136.

[57] Hemmo 2006, p. 88.

[58] Koulu 2016, p. 65.

to obtain the vending machine and a location for it, set up the vending machine and fill it with products, programme the vending machine and make it operational before any contracts can be concluded. The user also expresses their will to be bound to the transaction similarly via the vending machine. The vending machine example can also be described using the offer–acceptance mechanism; however, tacit agreements seem more relatable to the reality of the phenomenon.[59]

The Supreme Court of Finland has stated in case KKO 2010:23 regarding private parking enforcement that the offer–acceptance mechanism of the Contracts Act no longer corresponds with all situations related to the conclusion of a contract. Contracts concluded via automats were mentioned in the ruling as another relevant example of these types of contracts.[60] The conclusion of a contract can therefore also be attributed to external characteristics presented in the parties' actions.[61]

Conclusion of a Smart Contract

In the previous section, we presented a number of mechanisms for concluding a contract. In this section, we will be comparing these mechanisms and evaluating how well contract law doctrines regarding the conclusion of contracts are applicable to smart contracts.[62]

Especially in the offer–acceptance mechanism of the Contracts Act, the parties' declarations of intent are explicit; in other words the acceptor is given the details of the offer and the offeror is given information on the response. On the other hand, as explained previously, consent can be expressed implicitly, for instance, through co-operation with the other party or the performance of duties. Since the doctrine on declaration of intent holds a strong principal position in the Finnish legal system, this must also be taken into account when discussing the conclusion of a contract from the perspective of smart contracts.

In reference to what has been discussed previously, it appears possible that smart contracts can be concluded based on the parties' declaration of intent. Although it

---

[59] Saarnilehto et al. 2012, pp. 384–385.

[60] KKO 2010:23: "*Esimerkkeinä sopimuksista, joiden syntymisen edellytysten tarkasteluun oikeustoimilain periaatteet tuntuvat riittämättömiltä, on usein mainittu muun muassa erilaisia teknisiä välineitä, kuten automaatteja hyväksi käyttäen tehdyt sopimukset sekä sellaiset sopimukset, joita tehdään päivittäin ja toistuvasti suuria määriä ja jotka keskeiseltä sisällöltään ovat aina samanlaisia [...]*". [As examples of contracts, the conclusion of which the principles of the Contracts Act seem insufficient to explain, two similar contract types can be mentioned: contracts concluded using various technical devices, such as automats, and contracts concluded again and again in large quantities which are essentially always the same by content.]

[61] Saarnilehto et al. 2012, pp. 384–385.

[62] This may also be interesting in order to evaluate the effects on third parties, i.e. *ultra partes*. Even though the matter will not be discussed further in this text, it contains very important follow-up questions outside of contract law, e.g. in relation to tort liability, consumer protection, jurisdiction, conflicts of laws as well as dispute resolution.

seems that the offer–acceptance mechanism can be applied to smart contracts, their conclusion seems to be better explained by the processes leading to tacit agreements and implied contracts. In the context of the offer–acceptance mechanism, the parties would come to a binding agreement via the offer of one party and the acceptance of the other. Only thereafter are transactions or other actions performed in accordance with the contract. With smart contracts, the intent of the party responsible for placing the smart contract in the blockchain seems to manifest in the same context where a contracting party transfers a certain digital asset to be managed by the smart contract.[63] Declaration of intent does not therefore appear to occur separately from the conclusion or execution of a smart contract but is rather an immovable part of the contract itself.[64] Then again, if observed in light of the offer–acceptance mechanism, a public smart contract added to the blockchain to which the party has transferred assets for management may perhaps be interpreted as an offer.[65] Respectively, another party's joining the smart contract may be seen as acceptance of the offer.[66]

The expressions of intent in the conclusion of a smart contract share many characteristics with a tacit agreement, where the contract is concluded by parties exchanging assets. When a party transfers the sum into the smart contract, and the other party begins to act based on the smart contract, the expressions of intent of both parties are included in the actions taken. Even though no deliberate expressions of intent are given, the actions of the other party are required in order to be bound to the contract.[67] A parallel can be drawn between this situation and the previously mentioned situation involving an automat. This interpretation is enforced partly by the fact that Szabo has mentioned in some of the first publications about smart contracts that an automat is the simplest form of a smart contract.[68]

---

[63] This manner of concluding a contract includes some similarities to the aforementioned real contracts. While real contracts often require the subject matter of the contract to be lodged in the custody of the other party, it would have to be separately evaluated to what extent the transferred sum controlled by a smart contract could constitute such a subject matter.

[64] Koulu 2016, p. 65: "*The declaration of intent is not separate from the formation of the contract or from the execution of it*".

[65] It is a question of its own whether this type of offer and its acceptance are precise enough to meet the requirements of the offer–acceptance mechanism. When an announcement alone that a party is willing to conclude contracts does not necessarily constitute an offer (but rather an invitation to make one), the smart contract in the blockchain might not be such a specific offer either. See, e.g. Saarnilehto 2009, pp. 42–43.

[66] What may become interesting is the type of situation in which a complex smart contract has a wide range of unspecified creators, where it may be impossible to identify the offering party. A compelling question here is for instance how a group like this can validly act as an offeror. This theme will not, however, be discussed any more widely in this article.

[67] A different interpretation could be formed in a situation where it would be possible to commit to a smart contract by mistake or without understanding its true code-form content. These types of situations may be possible as the use of smart contracts becomes more popular, and it will be important to observe these situations in the future.

[68] Despite previous evaluations, a smart contract is not, for example, a mechanical automat containing beverages, but rather a programme which performs a specified action based on its programmed execution logic. A nearly infinite amount of different kinds of smart contracts can be programmed,

Based on aforementioned details, acts performed by the parties of a smart contract can likely be thought to fulfil the definition of declaration of intent.[69] Therefore, at least certain types of smart contracts can feasibly be concluded either by acting upon them or implicitly, as demonstrated in the aforementioned vending machine example. Here the "creator" of the smart contract announces their will to conclude contracts by building a smart contract in the blockchain and transferring, for example, certain assets to it. The other party of the smart contract expresses their will to be bound by performing an act in accordance with the terms of the contract, therefore accepting the offer without a distinct and explicit declaration of intent. Finally, when the preconditions specified in the smart contract are met, it executes itself automatically and, for example, redistributes the digital assets placed under its management or performs other tasks it has been appointed with, following which the contract can be thought to have been expired.[70]

However, not all smart contracts are as simple in reality. Next, we will discuss examples of different types of smart contracts and aim to highlight their various characteristics.

Can Smart Contracts Be Used to Perform Legal Acts?

Case: API Oracle

The first example is about so-called oracles, in other words routers connecting a set of application programming interfaces (APIs). This type of smart contract collects data from one or more third-party software interfaces or other sources and relays the collected information into the blockchain.

The main purpose of oracles is to provide information to other smart contracts in order to monitor the fulfilment of the terms of the contract. This is to ensure that one of the basic requirements of a functional consensus architecture is met: each party must be able to check the validity of the information in the blockchain. If the smart contracts were to monitor the fulfilment of the terms of the contract via information available on typical websites or third-party software interfaces, then the risk would be that each party would find different results, thereby undermining the reliability of the contracts. Hence all factors which will affect the smart contracts must be brought into the blockchain through oracles.

---

so it is quite probable that not all smart contracts can be seen to involve the type of (at least implied) declaration of intent that is required to conclude a legally relevant act.

[69] In this chapter we have discussed smart contracts in accordance with the definitions described previously in this publication. In addition, it has been considered that a smart contract only has one creator and is joined by only one other party.

[70] The true intelligence of smart contracts can be questioned, as they do not contain artificial intelligence in themselves, as has been stated previously in this publication. A smart contract should thus be perceived as an automated mechanism which performs its defined functions as certain preconditions are met. The established term "smart contracts" is thus somewhat deceiving.

Quite understandably, there are some trust issues related to using individual oracles, where one wants to maintain the benefits of using decentralised consensus architecture. In its simplest form, however, a smart contract functioning as an oracle would appear as follows:

```
pragma solidity ^0.4.11;


contract Oracle {


    address oracle;
    uint[] public data;

    function Oracle() {
        oracle = msg.sender;
    }

    function reportData(uint newData)
    {
        require(msg.sender == oracle);
        data.push(newData);
    }
}
```

Obviously the oracle in itself does not resemble what is commonly understood in our contract law as a contract. The example given above contains no typical features of a contract. In addition, the smart contract does not include identifiable parties and therefore does not include anyone's expression of intent. Its only purpose is to collect data from one location and send it to another. This type of a smart contract functions specifically as a programme designed to relay data. This example quite clearly illustrates the problems caused by the discrepancies between the terminology and contents of smart contracts. Even though the entirety of the contracts which the oracle is a part of may resemble a typical contract, the oracle in itself would still be nothing more than a programme designed to relay data.

Case: Search Engine Optimisation

A slightly different example of a smart contract is a basic service level agreement. This type of contract could, for example, be used to estimate the success of search engine optimisation. In this scenario, a buyer looking to purchase search engine optimisation services has created a smart contract into a blockchain, specifying the optimisation services required. The buyer will deposit the offered amount of value into the contract. A seller who wishes to enter into the agreement does so by also depositing an amount of value into the contract as collateral. Once the deadline specified in the terms of the smart contract is due, the contract will assess whether the buyer's domain is amongst the top three Google search results for the search term "example", conducted by a specified oracle. If the terms of the contract are met at the time of the deadline, the seller will receive both of the deposited sums. Conversely, if the terms are not met, both of the deposits will go to the buyer. The described smart contract could be written as follows:

```
    pragma solidity ^0.4.11;


    contract GoogleSearchOracle {
       function getRanking(string url, string searchTerm) constant
returns (uint);
    }

    contract ServiceLevelAgreement {


        GoogleSearchOracle oracle = GoogleSearchOracle(0x8b208798
4b3b3f15450a644887f100d9559bb0cc);


        address buyer;
        address seller;

        uint price = 190 ether;
        uint collateral = 2 ether;
        uint maxAcceptedRank = 3;
        string domainName = "http://www.example.com/";
        string searchTerm = "example";
        // 2017-10-15 at 0 hours 0 minutes 0 seconds in Unix time
        uint deadline = 1508025600;

        function ServiceLevelAgreement() payable {
            require(msg.value == price);

            buyer = msg.sender;
        }

        // The contract can be canceled as long as it hasn't been
        // accepted by anyone
        function cancel() {
```

```
            require(msg.sender == buyer);
            require(!seller);

            selfdestruct(buyer);
        }

        function accept() payable {
            require(!seller);
            require(msg.value == collateral);

            seller = msg.sender;
        }

        function doSettlement() {
            require(seller);
            require(now >= deadline);

            // By default, send the deposit to the seller of the
            // service...
            address recipient = seller;
           // ...but, if failed to reach the agreed service level,
return the deposit to the buyer
           if (oracle.getRanking(domainName, searchTerm) > maxAc-
ceptedRank) {
                recipient = buyer;
            }

            if (!recipient.send(price + collateral)) {
                throw;
            }
        }
}
```

In this example, the buyer has drafted a contract-like digital instrument and deployed it in a public blockchain. This act can be interpreted as an indication of the buyer's willingness to enter into an agreement. The seller demonstrates the same willingness to enter into an agreement by depositing the predetermined sum of value into the contract. Such a construction is very similar to a tacit agreement and is therefore quite a clear example of how legal acts can be performed with smart contracts.

It is noteworthy, however, that although the smart contract in this example allows the contracting parties to align their incentives in such a way as to achieve their contractual goals, technically the arrangement itself does not involve any contractual obligations for the seller to optimise the search engine results. Essentially the contract constitutes a simple bet on the search result placement of a certain domain on a given date, at a given time. It simply then follows from this bet that the passivity of the seller in this respect would result in the loss of the seller's own deposit and the forfeiting of the buyer's deposit.[71]

Based on this example, when evaluating the legal position of smart contracts and the obligations and rights which they create for the parties involved, it bears significance how and between which parties the smart contract was created.[72] In light of our current legislation dealing with contract law, the casuistic nature of the evaluation is emphasised.

Case: Token Sale (a.k.a. Initial Coin Offering, ICO)

Smart contracts can also be used for purchasing shares in so-called token sales, or *initial coin offerings (ICO)*. The idea herein is somewhat analogous to crowdfunding applied to pre-seed venture capital funding rounds for start-ups. As funds are paid into the smart contract, tokens are transferred to the purchasing party to represent the ownership of shares. These tokens can be programmed to include several types of functionality, including dividends, voting rights and access to goods and services later on produced by the company.[73]

In this example, in order to raise funds for a start-up company, an issuer is offering to sell share-representing tokens for a predetermined price of 1 ether per token and offers to accept all purchases conducted before the set deadline. The smart contract could be drafted as follows:

[71] Another perspective to the smart contract in this example is that of contractual penalties. It could be interpreted that the deposit required from the seller in order to enter the agreement constitutes a contractual penalty clause.

[72] Regarding the example, declaration of intent may manifest in different ways within the scope of the applied conclusion mechanism, depending on which party is the creator of the smart contract and which party is the one reacting to the smart contract. If a party of the arrangement does not act as the creator of the smart contract or react to it by making a payment or digital signature, their declaration of intent may be very difficult to prove.

[73] For further information on ICOs, see e.g. Conley 2017.

```
pragma solidity ^0.4.11;


contract ICO {


    address tokenIssuer;
    uint collectedEther;
    uint minFunding = 2500 ether;
    mapping (address => uint) public balances;
    // 2017-10-15 at 0 hours 0 minutes 0 seconds in Unix time
    uint icoDeadline = 1508025600;

    function ICO() {
        tokenIssuer = msg.sender;
    }

    function mint() payable {
        require(now < icoDeadline);

        collectedEther += msg.value;
        balances[msg.sender] += msg.value;
    }

    function transfer(address receiver, uint amount) {
        require(fundingSuccessful());
        require(balances[msg.sender] >= amount)

        balances[msg.sender] -= amount;
        balances[receiver] += amount;
    }

      // If funding was successful, the token issuer may
withdraw
    // all deposits
    function withdrawFunding() {
        require(fundingSuccessful());

        if (!tokenIssuer.send(collectedEther)) {
            throw;
        }
    }

    // If funding failed, investors may withdraw their
    // investments back
```

```
        function withdrawInvestment() {
                require(now >= icoDeadline && collectedEther <
minFunding);

            uint investment = balances[msg.sender];
            balances[msg.sender] = 0;

            if (!msg.sender.send(investment)) {
                throw;
            }
        }

           function fundingSuccessful() private constant returns
(bool) {
           if (now >= icoDeadline && collectedEther >= minFunding)
{
                return true;
            }
            return false;
        }

        function payDividends() {
            ...
        }

        function vote() {
            ...
        }
    }
```

In this case, the issuer of the token sale has drafted a smart contract and publicly deployed it, specifying the offered price, the minimum funding threshold and the termination deadline of the offer. Investors wanting to engage in an investment arrangement with the issuer can do so by transferring their stake as cryptocurrency tokens into the smart contract. Once the termination deadline has been reached and the offer has expired, the smart contract will determine whether a sufficient amount of funds has been committed to the funding round. If the minimum threshold has been surpassed, the contract will release the funds transferred into the contract to the issuer of the token sale, and the funders will be issued share-representing tokens accordingly.

In this example, the expressions of intent of the parties are quite clear, and the contract can be seen to have been concluded tacitly. The issuer's expression of intent (offer) is manifested in the act of deploying the smart contract into a blockchain, and the funder's reciprocal acceptance takes form in the depositing of the funds into the smart contract. The situation can therefore be interpreted via the offer–acceptance mechanism found in the Finnish Contracts Act such that the issuer has shown their willingness to enter into the contract by placing the smart contract into a blockchain, and the funder has reciprocated by transferring the funds. If the offer has been sufficiently identifiable then this interpretation is viable. The third example seems to reinforce the understanding that a smart contract can be a contract in the typical legal sense of the word, if an offer–acceptance mechanism can be sufficiently identified. This view is further reinforced when the example is interpreted analogously in comparison with the vending machine example.[74]

Conclusions and Discussion

Smart contracts can be drafted on very different bases and for entirely dissimilar purposes – not all of which meet the characteristics and the legal requirements of a contract. Based on the empirics in section "Can Smart Contracts Be Used to Perform Legal Acts?", however, it seems rather clear that legal acts *can* be concluded in the form of smart contracts. In this regard, the manifestation of intent through the exchange of performances appears to be of focal importance. A similar mechanism has been previously presented in the Finnish legal literature – namely, the vending machine, where the implicit nature of declarations of intent is highlighted in the formation of the contract. However, due to the fact that smart contracts are not specifically covered in the current legislation, legal ambiguity may arise as a consequence of their conceptual unconventionality.[75]

In addition to the ambiguity in regard to the letter of the law, smart contracts can also be subject to algorithmic ambiguity, so to speak. When co-operation is organised just by the programming code of a smart contract, trying to understand the true legal content of the arrangement on the basis of the programming code alone can be problematic.[76] Although this chapter has described three examples of smart contracts, in reality the number and the scope of possible applications may be practically

[74] This type of a smart contract seems to include characteristics of a contract containing conditions precedent or subsequent. In so-called conditional sales, it can be agreed that the sale is only concluded if a certain future event takes place. Conditions subsequent refer to uncertain events. In this case the condition subsequent would manifest as the cancellation of the sale (and the return of the deposit to Y) in case the ICO fails to attract sufficient amounts of funding. For more about the conditions of a contract, see, e.g. Saarnilehto et al. 2012, pp. 401–402. Conditions and conditional sales will not be further discussed in this publication.

[75] For a similar interpretation from the Estonian perspective, see also Kolvart et al. 2016, p. 145.

[76] Conversely, however, it is worth noting that if a traditional contract were to be created in code, this would require the contract to be arranged and presented as a process depicting interdependency: "if X, then Y, otherwise Z" (Mattila 2016, p. 15). Since the way in which traditional contracts are worded can often result in ambiguity, this new use of formulas can in at least some cases reduce the need for interpretation (Wright and De Filippi 2015, pp. 11 and 24–25). This kind of development can at best lead to significant reductions in the costs caused by drafting contracts and overseeing their execution.

infinite. The variety of smart contracts may cause various legal issues, the effects of which may be hard to anticipate at such an early stage.[77]

With the focus on such potential challenges, "soft law" arrangements, such as so-called dual integration systems[78] and systems based on various model agreements,[79] have already been developed to help prove the existence of a contract in the legal domain.[80] It is thus likely that smart contracts will first and foremost be utilised in the context of standard-form contracts and other kinds of simple contracts that do not involve ambiguous legal terms. Nevertheless, engaging in discussions about developing the legal doctrinal composition of smart contracts, both on the national as well as the European Union level, should be considered an equally important and topical approach in the matter.

In the literature on platform economy, boundary resources have traditionally been understood as technical tools used to lower the threshold for third parties to join part of a company's platform ecosystem. The perspective of technical tools, however, has yet to be applied to social boundary resources on a similar scale. Smart contracts are a clear example of how social boundary resources are developing in an increasingly technical direction. It is becoming increasingly difficult to draw a distinction between technical and social boundary resources of platforms. Social boundary resources should therefore be perceived as technical enablers, similarly to technical boundary resources.[81]

Contracts in themselves have also not been formerly perceived as boundary resources, in the sense that the network effects of a platform ecosystem could be boosted by opening up so-called *application contracting interfaces, ACIs* (cf. *application programming interfaces, APIs*). This would enable the creation of more

---

[77] Such questions may regard, for instance, the existence of a contract or the verification of its content (code vs the parties' true intent) as well as possible unintended errors left in the code. For such errors related to the intent of the parties, it is likely that section 32(1) (concerning the so-called error in declaration) of the Finnish Contracts Act can be applied if there is a conflict between content and intent due to an error in the contract code. See, e.g. Hemmo 2003, p. 396.

[78] Dual integration: "The idea of dual integration is to allow users to be able to have the certainty of having a real world contract which can be taken to a court and enforced using established dispute resolution processes in the jurisdiction(s) of the user(s) while also using a smart contract as the primary mechanism for administering the data-driven interaction which attends to the agreement between the parties" (https://erisindustries.com/components/erislegal/ (23 August 2016).

[79] Out of these openly developed solutions, the perhaps most significant one is Common Accord: "[…] an initiative to create global codes of legal transacting by codifying and automating legal documents, including contracts, permits, organisational documents and consents. We anticipate that there will be codes for each jurisdiction, in each language. For international dealings and coordination, there will be at least one 'global' code". Well-known lawyer and crypto-oriented legal researcher Primavera De Filippi is part of the Common Accord group. See http://www.commonaccord.org/ (23 August 2016).

[80] One way to solve possible issues is by aiming to create general conditions of contract such as INCOTERMS or Creative Commons for the use of smart contracts. One such example is the Simple Agreement for Future Tokens (SAFT) initiative which aims to design a legally sound framework for carrying out initial coin offerings in accordance with the US legislation. See https://saftproject.com/ (5.12.2017).

[81] For comparison: Gawer 2009; Ghazawneh 2012; Ghazawneh and Henfridsson 2013.

highly automated digital contracting mechanisms, process automation that reaches further beyond companies' own information systems, as well as more automated and more dynamic networks of contracting parties.

In general, smart contracts can be expected to disrupt the development of the platform economy by enabling unprecedented ways to co-operate in open platform ecosystems. As for managerial implications, companies should address the following three considerations:

1. How can smart contracts be used to lower the threshold for third parties to enter the company's platform ecosystem, in the same manner as technical boundary resources have been used for opening interfaces and offering ready-to-go tools for development?
2. In cases where companies have several contracting interfaces towards their clients, suppliers and other interest groups, which interfaces are suitable for the use of smart contracts with each respective party?
3. If several parties are subjected to the same smart contract in a vending-machine-like manner, are contractual arrangements required by successful business strategy becoming more fragmented, if individual deliveries are comprised of several constituent parts of separate suppliers?

Acknowledgements This paper has been drafted as a part of the Work and Wealth in the Era of Digital Platforms (BRIE-ETLA) research collaboration of 2015–2018. Timo Seppälä has also received financial support for his work from the Integrating Platform Competences towards Network Effects (iPlate) research project.

References

Official References

Report of the Commission for reform of the Contracts Act, 1990:20.
Contracts Act, 228/1929.

Legal Praxis

KKO 2010:23.

Legal Literature

Hemmo, M. (2003). *Sopimusoikeus I*. Helsinki: Talentum Media Oy.
Hemmo, M. (2006). *Sopimusoikeuden oppikirja*. Helsinki: Talentum Media Oy.

Hirvonen, A. (2011). *Mitkä metodit? Opas oikeustieteen metodologiaan* (Vol. 17). Helsinki: Yleisen oikeustieteen julkaisuja.
Norros, O. (2007). *Vastuu sopimusketjussa*. Vantaa: WSOYpro.
Saarnilehto, A. (2009). *Sopimusoikeuden perusteet*. Helsinki: Talentum Media Oy.
Saarnilehto, Ari – Hemmo, Mika – Kartio, Leena (2012): Varallisuusoikeus; Sanoma Pro Oy, Helsinki.

Other Literature and Other Publications

Ailisto, H., Mäntylä, M., & Seppälä, T. (Eds.). (2015). *Finland—The Silicon Valley of industrial internet*. Publications of the Government's analysis, assessment and research activities No. 10.
Baldwin, C. Y., & Woodard, C. J. (2009). The architecture of platforms: A unified view. In A. Gawer (Ed.), *Platforms, markets and innovation* (pp. 19–44). Cheltenham: Edward Elgar Publishing.
Boudreau, K. J., & Hagiu, A. (2008). Platform rules: Multi-sided platforms as regulators. In A. Gawer (Ed.), *Platforms, markets and innovation* (pp. 1–29). Cheltenham: Edward Elgar Publishing. Available at: https://doi.org/10.2139/ssrn.1269966.
Bourque, S., & Fung Ling Tsui, S. (2014). *A Lawyer's introduction to smart contracts*. Łask: Scientia Nobilitat.
Buterin, V. (2013). *A next-generation smart contract and decentralized application platform* (Ethereum White Paper).
Buterin, V. (2014, May 6). *DAOs, DACs, DAs and More: An incomplete terminology guide*. Ethereum blog. Available at: https://blog.ethereum.org/2014/05/06/daos-dacs-das-and-more-an-incomplete-terminologyguide/
Chander, A. (2014). How law made Silicon Valley. *Emory Law Journal, 63*(3), 39–694.
Conley, J. (2017). *Blockchain and the economics of crypto-tokens and initial coin offerings* (Vanderbilt University Department of Economics Working Papers 17–00008).
Cusumano, M., & Yoffie, B. (1998). *Competing on Internet time: Lessons from Netscape and its battle with Microsoft*. New York: Free Press.
Cusumano, M. (2005). Google: What it is and what it is not. *Technology Strategy and Management, 48*(2), 15–17.
Cusumano, M. (2010). Technology strategy and management: The evolution of platform thinking. *Communications of the ACM, 53*(1), 32–34. Available at: http://cacm.acm.org/magazines/2010/1/55758-the-evolution-of-platform-thinking/abstract.
Cusumano, M. A., & Gawer, A. (2002). Platform leadership. *MIT Sloan Management Review, 43*(3), 51–58.
Eisenmann, T., Parker, G., & Van Alstyne, M. (2006). Strategies for two- sided markets. *Harvard Business Review, 84*(10), 1–10.
Gawer, A., & Cusumano, M. A. (2002). *Platform leadership*. Boston: Harvard Business School Press.
Gawer, A., & Cusumano, M. A. (2008). How companies become platform leaders. *MIT Sloan Management Review, 49*(Winter), 27–35.
Gawer, A. (2009). *Platforms, markets and innovation*. Cheltenham: Edward Elgar Publishing Inc.
Gawer, A., & Henderson, R. (2007). Platform owner entry and innovation in complementary markets: Evidence from Intel. *Journal of Economics and Management Strategy, 16*(1), 1–34.
Ghazawneh, A. (2012). *Towards a boundary resources theory of software platforms* (JIBS Dissertation Series No. 085).

Ghazawneh, A., & Henfridsson, O. (2013). Balancing platform control and external contribution in third-party development: The boundary resources model. *Information Systems Journal, 23*, 173–192.
Glatz, F. (2014). What are smart contracts? In search of a consensus. *Medium.com*.
Hagiu, A. (2014). Strategic decisions for multisided platforms. *MIT Sloan Management Review, 55*(2), 71–80.
Kenney, M., & Pon, B. (2011). Structuring the smartphone industry: Is the mobile internet OS platform the key? *Journal of Industry, Trade and Competition, 11*, 239–261.
Koulu, R. (2016). Blockchains and online dispute resolution: Smart contracts as an alternative to enforcement. *ScriptEd, 13*(1).
Kõlvart, M., Poola, M., & Rull, A. (2016). Smart contracts. In T. Kerikmäe & A. Rull (Eds.), *The future of law and eTechnologies* (pp. 133–147). Cham: Springer. Available at: https://doi.org/10.1007/978-3-319-26896-5_7.
Mattila, J. (2016). *The blockchain phenomenon – The disruptive potential of distributed consensus architectures* (ETLA Working Papers No. 38).
Mattila, J., & Seppälä, T. (2015). *Blockchains as a path to a network of systems: An emerging new trend of the digital platforms in industry and society* (ETLA Reports 45).
Mattila, J., Seppälä, T., & Holmström, J. (2016). *Product-centric information management – A case study of a shared platform with blockchain technology* (Conference paper). Industry Studies Association Conference 2016.
Nakamoto, S. (2008). *Bitcoin: A peer-to-peer electronic cash system*. Available at: https://bitcoin.org/bitcoin.pdf
Narayanan, A., Bonneau, J., Felten, E., Miller, A., & Goldfeder, S. (2016, February 9). *Bitcoin and cryptocurrency technologies* (Draft).
Parker, G., & Van Alstyne, M. W. (2014). *Innovation, openness & platform control*. Available at: https://ssrn.com/abstract=1079712
Parker, G. G., & Van Alstyne, M. W. (2005). Two-sided network effects: A theory of information product design. *Management Science, 51*(10), 1494–1504. Available at: http://pubsonline.informs.org/doi/abs/10.1287/mnsc.1050.0400.
Parker, G. G., Van Alstyne, M. W., & Choudary, S. P. (2016). *Platform revolution: How networked markets are transforming the economy and how to make them work for you*. New York: WW Norton & Company.
Pon, B., Seppälä, T., & Kenney, M. (2014). Android and the demise of operating system-based power: Firm strategy and platform control in the post-PC world. *Telecommunications Policy, 38*(11), 979–991.
Pon, B., Seppälä, T., & Kenney, M. (2015). One ring to unite them all. Convergence, the smartphone, and the cloud. *Journal of Industry, Competition and Trade, 15*(1), 21–33.
Scholz, L. (2016, forthcoming). *Algorithmic contracts stanford technology law review*.
Seppälä, T., Halén, M., Juhanko, J., Korhonen, H., Mattila, J., Parviainen, P., Talvitie, J., Ailisto, H., Hyytinen, K.-M., Kääriäinen, J., Mäntylä, M., & Ruutu, S. (2015). *"Platform" – Historiaa, ominaispiirteitä ja määritelmä* (ETLA Raportit – Reports 47).
Szabo, N. (1994). *Smart contracts*. http://www.fon.hum.uva.nl/rob/Courses/InformationInSpeech/CDROM/Literature/LOTwinterschool2006/szabo.best.vwh.net/smart.contracts.html. Accessed on 15 June 2017.
Szabo, N. (1996). *Smart contracts: Building blocks for digital markets*. http://www.fon.hum.uva.nl/rob/Courses/InformationInSpeech/CDROM/Literature/LOTwinterschool2006/szabo.best.vwh.net/smart_contracts_2.html. Accessed on 15 June 2017.
Szabo, N. (1997). *The idea of smart contracts*. http://www.fon.hum.uva.nl/rob/Courses/InformationInSpeech/CDROM/Literature/LOTwinterschool2006/szabo.best.vwh.net/idea.html. Accessed on 15 June 2017.

Szabo, N. (2014). *The dawn of trustworthy computing*. http://unenumerated.blogspot.fi/2014/12/the-dawn-of-trustworthy-computing.html. Accessed on 15 June 2017.

Tapscott, D., & Tapscott, A. (2016). *Blockchain revolution: How the technology behind bitcoin is changing money, business and the world*. London: Portfolio Penguin.

Tiwana, A., Konsynski, B., & Bush, A. (2010). Platform evolution: Coevolution of platform architecture, governance, and environmental dynamics. *Information Systems Research, 21*, 685–687.

Werbach, K., & Cornell, N. (2017). Contracts Ex Machina. *Duke Law Journal* (Forthcoming), pp. 1–57.

Wright, A., & De Filippi, P. Decentralized Blockchain Technology and the Rise of Lex Cryptographia. (2015, March 10). Available at SSRN: https://ssrn.com/abstract=2580664 or https://doi.org/10.2139/ssrn.2580664

Yoo, Y., Henfridsson, O., & Lyytinen, K. (2010). The new organizing logic of digital innovation: An agenda for information systems research. *Information Systems Research, 21*(4), 724–735.

Open Access This chapter is licensed under the terms of the Creative Commons Attribution 4.0 International License (http://creativecommons.org/licenses/by/4.0/), which permits use, sharing, adaptation, distribution and reproduction in any medium or format, as long as you give appropriate credit to the original author(s) and the source, provide a link to the Creative Commons license and indicate if changes were made.

The images or other third party material in this chapter are included in the chapter's Creative Commons license, unless indicated otherwise in a credit line to the material. If material is not included in the chapter's Creative Commons license and your intended use is not permitted by statutory regulation or exceeds the permitted use, you will need to obtain permission directly from the copyright holder.

Digital Platforms for Restructuring the Public Sector

Antti Hautamäki and Kaisa Oksanen

Abstract Many technological innovations have led to the emergence of the platform economy in recent years. This development is changing the entire landscape of business in the era of digitalisation. However, the impacts of the platform economy on public services and government are not well known. In this article we study the potential for the digital platform economy to help restructure the public sector. Firstly, central features of the new platform technology are explored, pointing to an algorithmic revolution, big data and cloud computing. Platforms are used in coordinating market transactions in an extremely efficient way. In order to apply the platform concept to the public sector, an experimental approach is needed; public platforms cannot be built by transposing mechanical models of the private sector to the public sector, because the market logic of public services is quite different than open markets. To illustrate the challenges and possibilities of the platform economy, we explored a few cases from Finland such as "Suomi.fi" digital service platform and its background technology, which is based on a national architecture for digital services developed in Finland applying X-Road technology created originally in Estonia. As a special case, we studied the Finnish solution to the digital health-care system. The case of "Kanta Services" exemplifies the challenge to simultaneously develop open and secure data systems for health care. Finally, we point out the importance of citizen-centred approaches in developing platforms for the public sector.

Keywords Platform economy · Public services · Restructuring the public sector · Artificial intelligence · Mydata

---

A. Hautamäki (✉)
University of Jyväskylä, Jyväskylä, Finland

K. Oksanen
Prime Minister's Office, Helsinki, Finland
e-mail: kaisa.oksanen@vnk.fi

© Springer Nature Singapore Pte Ltd. 2018

A. Smedlund et al. (eds.), *Collaborative Value Co-creation in the Platform Economy*, Translational Systems Sciences 11,
https://doi.org/10.1007/978-981-10-8956-5_5

Introduction

In recent years, information and communication technology has taken considerable leaps towards a digital revolution of service systems in the public sector as well as in business (Zysman 2006). The issue is not only to reform services by applying digital technology but even more to create entirely new services. This is based very much on the emergence of digital platforms as a new way to coordinate actions of a great number of actors in society. A platform generates and orchestrates a marketplace where supply and demand meet in a transparent and effective manner. At the same time, three other new features have opened new applications of information technology. They are artificial intelligence, big data and cloud computing. Cloud computing, in particular, is liberating service producers from extra investments in infrastructure and software. Artificial intelligence makes it possible to analyse data and big data in a way that over-performs human experts in many fields of expertise. These developments have made possible the emergence of the platform economy as a new phase of the economy.

In this article, we consider how the platform economy is changing public services and government. First we explain the need for restructuring government services. We point out that the platform economy might be the next big thing in public sector innovation. Then we explore the basic features of the platform economy, stressing how it will change the logic of business. We connect the platform economy to other recent developments of technology, especially cloud computing. We also consider the sharing economy as a form of the platform economy. It is important for the public sector, because it is based on citizens' own activity and thus is a form a democracy.

The majority of this article is devoted to addressing the question of how to benefit from the platform economy in the public sector. In particular, the possibilities for restructuring public services by platforms are clarified via the chosen case studies. We have selected the new Finnish architecture for public services as an important case, because it is a complicated system exemplifying many features of the platform economy. Suomi.fi is a comprehensive service platform for almost all digital services of the Finnish public service and government. For health care, Kanta service is another comprehensive service platform in use in Finland.

Health care is a promising application of the platform economy, but it also presents a very complex and challenging task to secure the personal and very sensitive data of clients. Many of the current health-care platform actors are doing business mainly in the self-monitoring, lifestyle or preventive health-care sectors, and not providing clinical patient care (see, e.g. Smedlund 2016). Therefore we analyse the issue of data ownership and stress the right of a citizen to own her own data (MyData principle). In the conclusion, we discuss the conditions for developing an effective and secure public sector using the platform as a central tool.

The Need for Restructuring Government Services

Reinventing the form of government has been a global trend for some time. Phenomena such as privatisation or decentralisation have been tried at all levels of government. The promises of government reforms usually include enhancing efficiency, cutting costs, delivering better outcomes and strengthening citizen choice. In practice, actions have often included steps such as simplifying the regulatory framework or introducing new commercial actors within the public policy fields. These changes are being driven by a variety of forces, including a more globalised and networked world, rising citizen expectations, new technologies, increasingly complex problems facing governments and – particularly since the 2008 economic crisis – tight budgets.

Governments are also seeking to innovate. Governments seek to innovate in how they work, in the services they provide and how they provide them and in how they interact with citizens, businesses and civil society. Whatever the reason, the consensus seems clear: public sector organisations need new ways of working (OECD 2015). The overall goal of government innovation is to deliver better outcomes, such as better use of public resources, more open and trusting societies and strengthened justice and care for citizens from all walks of life (OECD 2017).

At the same time, the challenges facing governments are more complex due to technological and cultural changes, demographic changes and the global movement of resources and people. Similarly, public sector innovation has several limitations; for example, there are substantial inherent structural barriers, limited investment for innovation and deeper cultural barriers blocking disruptive thinking. In addition, the open use of public data and knowledge remains challenging in many places. On top of this, the analysis of innovative government remains limited and fragmented. Harnessing creativity in the public sector requires developing a better understanding of what creates successful innovations where the mechanics of change and its enabling factors are understood, alongside an understanding of the particular challenges faced by the public sector and the needs and preferences of its users (IPP 2017). For example, the OECD has continuously called for a suitable framework and tools for measuring public sector innovation (IPP 2017).

The newest and currently most pressing questions for reforming governments have been analysed by the OECD Observatory of Public Sector Innovation (OECD 2017). These questions include how to make the most of technology, how to work with citizens and draw on the abilities of society at large to address needs and how to rapidly test new approaches and ways of working in a fast-changing world.

In answering these questions, the platform economy can play an important role. It can be said that the platform economy is currently and will continue to be the state-of-the-art in public sector innovation. Because platform economy disruption is generally seen as the beginning of something new, something different and something cutting edge, it is also anticipated that it can transform how we make social and political choices (Vazquez Sampere 2016; Kenney and Zysman 2016).

The ABCs of the Platform Economy: Algorithmic Revolution, Big Data and Cloud Computing

The recent development of information technology has created possibilities for totally new solutions to many problems, which were difficult to manage earlier. We refer here to digital platforms, which are extremely effective ways to connect different actors of society. More specifically, the core problem that needs to be solved is the coordination problem. By this we mean the problem of coordinating actions of many actors who do not know each other. Typically, we face the coordination problem in a market, where sellers and buyers try to find each other and perform a transaction. In time before information technology, the coordination problem was solved by organising geographic (local) marketplaces for meetings and transactions. In local markets, trust was created through personal acquaintances. Business transactions are increasingly taking place in virtual spaces. Platforms, however, are more than virtual marketplaces – the essence of platforms is in their ability to enhance the co-creation of value that results in systemic offering of products and services (Smedlund 2016). The conditions for trust are quite different in the platform economy than in a "meeting economy".

It is interesting that the coordination problem has not yet been solved in a satisfactory way. To improve trust, several security improvements have been proposed. The newest approach is blockchain technology used in the Bitcoin money (cf. Owen 2015). Blockchain is a distributed database that maintains a dynamic list of ordered records, "blocks".[1] Each block contains a timestamp and a link to a previous block. This architecture makes it difficult or even impossible to change blocks afterwards. What is important is that a blockchain database is managed autonomously, and there is no need for supervising it.

The development of the Internet, the growth of the calculating powers of computers and software innovations opened the way for emerging digital platforms. Platforms are "software-based products or services that serve as a foundation on which outside parties can build complementary products and services" (Tiwana 2014, 5). Software platforms provide the core functionality shared by apps that interoperate with them together with interfaces, which they interoperate. Thus platforms are places where end users can benefit from the offer of applications. They can be likened to department stores, where different brands offer their products and deliver them from a warehouse. Among physical products, department stores also usually provide services such as barbershops and spas.

It is important to distinguish between platforms and single service providers. In a platform there are many service providers using the same platform. Therefore an ecosystem can emerge around the platform. An ecosystem on a platform is a combination of the platform and apps that interoperate with it (Tiwana 2014, 6). In business, competition is now taking place between ecosystems. A noteworthy example would be the competition between Apple and Nokia, where Apple succeeded to

[1] https://en.wikipedia.org/wiki/Blockchain, accessed March 31, 2017.

create a viable ecosystem with hundreds of thousands of application providers and destroyed the business of the leading cell phone producers. The advanced technology of Apple phones was not the major reason for winning the battle. Instead it was the ecosystem approach.

In a platform economy, owners of platforms occupy a central position. They bridge end users and app providers, making transactions easy to perform. Platform owners build the infrastructure and develop software, an interface for an app developer to enter the platform. The architecture of platforms is new and based on cloud computing, meaning that app providers or end users are no longer in need to make their own investments of infrastructure, data storage or even software. The Apple Store is a good example of this kind of development.

In recent years, a new kind of economy has been developing alongside the platform economy. It is the sharing economy, which refers to peer-to-peer based sharing of goods and services. A good example of this is Uber, which connects car owners and people in need of local transportation. Quite often the term sharing economy is used in a more general sense meaning just using an online marketplace for selling and buying products and services (like Zipcar, see Sandararajan 2013). But then all online business turns out to be in the sharing economy.

The core of the sharing economy is to provide ordinary people an easy way to benefit from their assets like apartments, cars, sports equipment or even skills and knowledge. To be effective, digital platforms are needed for sharing. An important question is where does the profit come from and whom does it benefit? Platform owners could earn a slice from each transaction, and asset owners earn from rents and services. But it is important to note that a sharing economy includes also voluntary actions and collaborative consuming without direct business implications. In this kind of social sharing, platforms are provided by non-profit organisations (Gore 2014).

The platform economy is connected to the general development of information technology. Especially important is an algorithmic revolution, by which John Zysman (2014) means that "tasks underlying services can be transformed into formal, codifiable processes with clearly defined rules for their execution". In the algorithmic revolution, activities are formalised and codified, and therefore they become computable. An algorithmic revolution opens paths towards artificial intelligence: developing algorithms for analysing data and making decisions.

An extremely interesting idea is to combine big data and artificial intelligence. IBM's Watson intelligent system has been the most successful application of this model. The system is over-performing many experts in medicine and other fields of high expertise. With Watson, one can analyse and interpret all data, including unstructured text, images, audio and video, utilise machine learning and create chatbots.[2]

Almost all platforms use cloud computing that delivers computing services such as data storage, computation and networking. Users will get the services at the time, to the location and in the quantity they wish to consume, with costs based only on the resources used (Kushida et al. 2014).

---

[2] IBM. 2017. https://www.ibm.com/watson/, accessed March 31, 2017.

Cloud computing architecture has three layers:

I. Application: Software as a Service (SaaS), e.g. Google Docs
II. Platform: Platform as a Service (PaaS), e.g. Windows Azure
III. Infrastructure: Infrastructure as a Service (IaaS), e.g. Amazon Web Services

As a whole, cloud computing makes it possible that a service provider does not need to invest in extra resources for computing. Cloud computing is transforming computing from scarce to abundant resources (Kushida et al. 2014). Kushida et al. (2014) argue that cloud computing is becoming the fundamental infrastructure of the global economy.

In summary, there are many parallel trends in modern information technology that together lead to the emergence of the platform economy. These trends include:

A. Algorithmic revolution and artificial intelligence
B. Big data and data analytics
C. Cloud computing

This "ABC" combination is the background for our analysis of restructuring public sector and public services by platforms. Often the platform economy is considered a phenomenon of the private sector, not directly affecting the way public services are organised. To better understand the impact of the platform economy in restructuring public services and governance, we can consider changes the platform economy is causing. First of all, we have to consider the platform economy from the viewpoint of economics, not so much as a bundle of technological innovations.

From an economic perspective, platforms are "two-sided markets" or "multi-sided markets" that facilitate the exchange between different types of consumers that could not otherwise transact with each other (Gawer 2014). The attraction of using platforms is based on network effects. One group of agents benefits from the size of other groups that join the platform. The network effect is the dominant view in analysing the economics of platforms. Sometimes the network effect works so cumulatively that ultimately some platform or its ecosystem will win and "take all" (Eisenmann et al. 2006). Amazon is an example of such a winner-takes-all scenario; it is clearly dominating the market of online bookselling. Currently the platform economy is also re-engineering journalism and publishing; the convergence between journalism and platform companies was recently charted by Bell and Owen (2017). In the span of 20 years, journalism has experienced three significant changes in business and distribution models: the switch from analogue to digital, the rise of the social media and now the dominance of mobile and platforms. This last phase has seen large technology companies dominate the markets for attention and advertising and has forced news organisations to rethink their processes and structures (Bell and Owen 2017).

The sharing economy is growing rapidly. It shows how effectively a platform economy is creating new markets for small producers and service providers. To provide some examples, Etsy (etsy.com) is a New York-based platform for selling unique products made by private individuals. Etsy has 25 million items for sale, 1.7 million active sellers and 28.6 million active buyers. Etsy also offers a wide range of seller services and tools that help creative entrepreneurs start, manage and scale their businesses. Etsy's business is large-scale and far-reaching but operated by only a thousand employees.

Another example is Loconomics (loconomics.com), which is a San Francisco-based platform for local service providers. Loconomics is a worker-owned co-operative, using no middleman. The services provided by their platform include, e.g. home care, child care, pet care, self-care, transport and catering. Services are easy to book on a 24/7 basis, and payment is done via credit card. There is no commission and sellers have total control of their pricing.

The Platform Economy and the Public Sector

Providing a common ground for innovation, collaboration and ecosystem construction has been recommended in numerous public sector development projects and processes. In practice the recommendations refer to various platforms and platform tools that can facilitate collaboration within individual organisations, across government and with the public. Approaches such as common platforms that enable people to connect at a central location can impact the ability of organisations to join forces in developing innovative solutions to common problems and to scale innovation (Daglio et al. 2014; OECD 2017).

Developing public sector platforms is a part of the digitalisation of public service delivery. As the public sector is the world's largest service provider (PwC 2007), the development of the platform economy provides public sector means for transitioning towards better digital services. Generally, the goal is a deeper interaction between citizens and the state.

As the development of the platform economy is generally seen to increase the availability of globally produced services, it leads to rising expectations from the traditional public services and thus increasing the need for public sector platform development. Another stream in the development is the promotion for experimental culture within the public sector and the government (e.g. Annala et al. 2015).

Platforms for Experimental Government: Case Experimental Finland

Regarding experimental government, currently, one of the leading examples is found in Finland. One of the current government's (2015–) key projects in Finland is to promote an experimental culture. The aim is to find innovative ways to develop society and services, and the project falls under the scope of the priority area of digitalisation, experimentation and deregulation (Experimental Finland 2017). In 2017, Finland is launching an online platform to crowdsource and crowdfund citizen-driven innovation and experimentation, thus combining the elements of experimentation and platforms.

Including experiments and behavioural approaches into policy design is not a new thing – for example, the UK government has had the Behavioural Insights Team (BIT) using behavioural economics and psychology in policymaking since 2010 (BIT 2017). In the Finnish case of experimental government, the goals have been similar: to incorporate behavioural approaches into governmental steering practices and, in doing so, to make policies more user-orientated and efficient (Sousa Lourenço et al. 2016). Developing an experimental culture can also lower cultural barriers for public sector innovation. As the OECD notes (IPP 2017), the political context of public sector organisations, their highly visible activities and potentially high consequences of failure can reinforce a culture of risk aversion. The culture of experimentation on the contrary can even encourage failing, or at least it can mitigate the fear of failure (Breckon 2015).

The experimental platform for citizens developed by the Prime Minister's Office in Finland aims to generate practical ideas on how to improve Finland and to develop the ideas into experimental proposals and scale the proposals if successful. This form of connecting with citizens engages people by giving them shared responsibility in the work and success of government (IPP 2017). The platform approach can also help in funding and diffusing the experiments. As a part of the well-regulated Finnish system, an experimenting platform provided by the public sector helps conduct the experiments in an ethical and sustainable manner.

Platforms for Digital Public Services: The Suomi.fi Case

As a case study, we analysed the digital platform developed for public services in Finland in recent years. From 2005 to 2010, a considerable number of digital services were developed in the Finnish public sector and government. It is estimated that the Finnish government now has about 700 electronic services, and combined with municipalities, the number is even higher. The problem with this burgeoning development has been its decentralised nature: public institutions and organisations are autonomous, and they have developed their services based on their own needs. In this situation, no one has been considering the compatibility of different information systems, and so the sharing of data between different services out of network is very difficult and often impossible. As a solution to this, the government started to create a national architecture for digital services. It is based on concepts such as platform as a service and infrastructure as a service.

It is interesting to compare the Finnish service infrastructure to the X-Road system of Estonia. Estonia has been very active in developing their X-Road system into a general platform for public and private services. X-Road is a service infrastructure, which connects different databases and opens access to information systems of different service providers. The Estonian interface to X-Road is the eesti.ee portal. With this portal, citizens can access their health records and vote in elections. In

Finland a similar platform has been in use since 2015 with the name *palveluväylä* ("Service Road"). It is a layer of data exchange on the Internet. End users do not have direct access to *palveluväylä*, because it is infrastructure consisting of many different service applications and systems.

> The main data exchange solution for the Finnish public sector organisations, Suomi.fi services, is based on X-Road technology. Suomi.fi *palveluväylä* was introduced in Finland as part of the programme implementing the National Architecture for Digital Services, and the public sector organisations have a statutory obligation to use it.[3]

Suomi.fi services are produced in the National Architecture for Digital Services programme. The programme is financed by the Ministry of Finance and operatively managed by the Population Register Centre. Next we will take a closer look at the Suomi.fi services and platform.

The public sector was established a long time ago in many countries to develop information portals to provide information about public services and public government. In Finland that portal is called Suomi.fi (meaning Finland.fi). In the beginning these information portals were only for one-way communication channels, but later they became interactive, as is also the case with Suomi.fi. The idea of Suomi.fi is expressed in its slogan "One address for citizen services" (https://www.suomi.fi/). Suomi.fi is the single access point to access public services in Finland. The majority of this platform consists of links to pages of different institutions and bureaus like ministries, Kela, museums, courts, etc. To find information, the content is divided by topics, the six most popular being: migration, teaching and education, family and social services, health and nutrition, work and pensions and taxation and financing. Suomi.fi has been in extensive test use in years 2014–2017, and officially it started in the beginning of the year 2018.

The most interesting part of the Suomi.fi platform is its e-services. On this platform, citizens and firms can establish e-transactions with authorities with the help of e-services and forms in Suomi.fi. A typical procedure involves selecting a form, filling it out and submitting it along with the application included. Then the authority will process the application, and finally the decision will be sent back in electronic format or post. The identification is completed by using online bank identifiers, a mobile certificate or a certificate card. Citizens can save the forms they have used in the My e-services application. Citizens can open their own account in which they can receive official decisions and notifications concerning the services that are linked to a citizen's account electronically instead of by post. Communications between services and users are encrypted against intervention by outside parties using SSL encryption.

Suomi.fi also contains a link to a general information website *Public Service Info,* which will guide users to the right public service providers. It is not for communication with authorities like submitting forms. In 2017, a service portal for enterprises, *Yrityssuomi.fi* ("EnterpriseFinland.fi"), was integrated into the Suomi.fi

---

[3] Ministry of Finance, Finland. 2017. Finland and Estonia set up a joint institute to develop X-Road technology, http://vm.fi/en/article/-/asset_publisher/suomi-ja-viro-perustavat-yhteisen-instituutin-kehittamaan-x-road-teknologiaa, accesses March 28, 2017.

site. Yrityssuomi.fi is a comprehensive site for many kinds of services important to businesses, like knowledge about legislation, taxation and funding possibilities. There an enterprise can fill out forms and send them to authorities. Yrityssuomi.fi is developed and updated by the Ministry of Employment and the Economy. Suomi.fi also contains a toolkit site for civil servants for official collaboration between authorities: *Suomi.fi/workspace* – information and services for authorities.

Platforms for Health Care: Case Kanta Services

For citizens, access to health care is one of the most fundamental services and rights. In this area, benefits from digitalisation are considerable. In Finland, a comprehensive service system, called *Kanta Services* ("Base Services"), was developed and has been in use since 2010. Kanta is the national data system services for health-care services, pharmacies and citizens. The services include the electronic prescription, Pharmaceutical Database, My Kanta pages and Patient Data Repository. It has two parts, one for citizens and one for professionals. The most used service so far is electronic prescriptions.

> An electronic prescription is a prescription for medicines issued and signed electronically by a doctor. It is entered into a centralized database called the Prescription Centre. The Prescription Centre register is controlled by Kela. The national Prescription Centre contains all electronic prescriptions and the dispensing records entered on them by pharmacies. Based on the information held in the Prescription Centre, any pharmacy can dispense your medicines.[4]

The Patient Data Repository is a service in which health-care units enter patient records from their own data systems in a secure manner. This data repository includes in 2017 data about 5.4 million persons. It offers citizens the opportunity to examine their own medical records on their computer and grants the right to health-care professionals to see them.

My Kanta (*omakanta*) is an online service for citizens where they can browse their own health records and their medication recorded by health-care services. So far about two million people have checked their health records there. In My Kanta service, a citizen can see her electronic prescriptions, records related to their own treatment, laboratory tests and X-ray examinations and health records of dependents under 10 years of age. In My Kanta service, one can also request a refill of a prescription, save her living will and organ donation forms and consent to or refuse the disclosure of her personal data.

To evaluate the national architecture of digital services in Finland, we have to remember that the programme to create this architecture was created quite recently, and Suomi.fi portal has been in use since 2015, although the testing started some years before. In any event, this architecture is a good example of the usefulness of a

[4]Electronic prescription in the Kanta service, http://www.kanta.fi/en/eresepti-esittely, accessed March 28, 2017

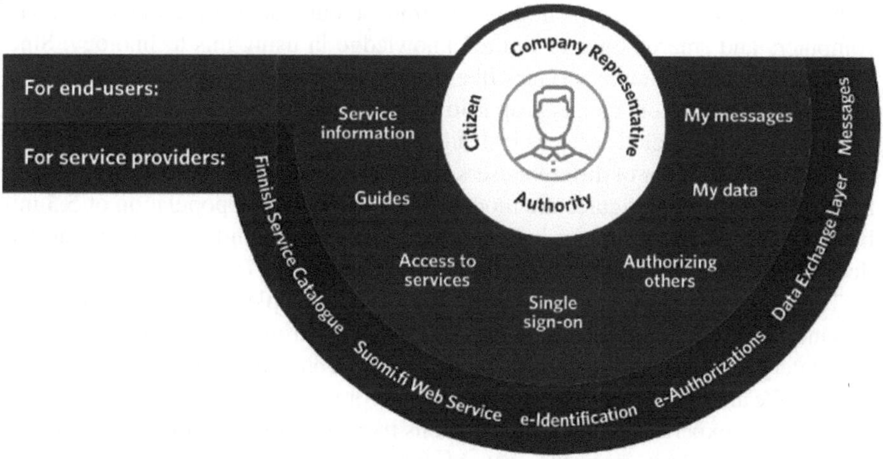

Fig. 1 Suomi.fi Services (Population Register Centre 6.3.2017, esuomi.fi)

new platform technology. The Suomi.fi system can be approached from two perspectives: from the point of view of the end user and of the service provider. In Fig. 1, both these perspectives are present. For end users, the amount and usefulness of web services are important as well as management of own data (MyData). For service providers, the service catalogue and services needed for building and running applications (identification, authorisations, data exchange, etc.) are critical factors. According to the Population Register Center, the basic structures for digital services are now established and are the most advanced in the world (Viskari 2017). Note however that the data exchange layer uses X-gate technology developed in Estonia.

The usefulness of Suomi.fi Services depends on the scope of the service. According to Henry Chesbrough (2011), the economies of scope mean lowering the average cost of a firm to produce two or more products. Although the cost of producing a service is not the first question asked by public organisations, still the incentives of benefiting from service provided by "joining" Suomi.fi portal might depend on the average cost. Public sector organisations have a "*statutory* obligation" to join Suomi.fi Services, like we quoted above. The issue is, however, that so many systems developed by autonomous public organisations are incompatible and expensive to convert. It takes years to renew basic information systems. Another problem is allowing a combination of data from different sources and registers. This is needed in order to guarantee the usefulness of public services (one interface for many services), but the data security and protection of identity are serious problems and challenging to solve.

Economies of scale refer to increasing the size of operations (Chesbrough 2011), which also relevant in evaluating Suomi.fi. The amount of operations or transactions is dependent on how citizens benefit from using services of Suomi.fi. Also, if the portal is difficult to access and use and provides no user support services, the danger is that many people will drop out. Here we think about older people who do not have computers and smartphones or limited knowledge in using this technology. Since the expenditures of public services like health care, family and social care, migration, work and pension, etc., is proportional to the success of e-services: how much citizens are using digital services instead of personal services. The data of Suomi.fi shows that the number of different users has grown from 2013 to 2015 from 176,788 to 242,502. Still these figures are modest in a country with a population of 5.5 million. But the situation is better in health-care systems: currently, with over one million visitors per month to My Kanta service (Ikävalko 2017).

There are two processes supporting the benefits of platforms (Hautamäki and Oksanen 2015). Commodification involves the move from special services towards elementary services, and scalability is the transfer from expert organisations to self-service. We are not saying that there is no need for special services like consultation with medical experts or expert organisations like specialised hospitals. The point is that to successfully manage the costs of public services, it is not possible without massive use of digital services by citizens. For this both economies of scale and scope are critical.

Finland has an outstanding opportunity to take a substantial leap in enhancing Suomi.fi services because the entire structure of health and social care systems is changing. Now the basic health care is provided by municipalities. The structure of health and social services will be reformed in a way that the responsibility for providing public health-care and social services will be assigned to 18 autonomous regions that are larger than municipalities.[5] This means among others that these 18 regional providers are much bigger than present over 300 municipalities. Another new feature is that public services might be generated by private companies and NGOs along with public service producers. Both of these reforms allow for coordinating digital service development and opening the Suomi.fi platform to private producers, thus enlarging the scope of the platform.

Artificial Intelligence and Big Data in the Public Sector

The platform economy is rapidly being adopted as a guiding principle for developing the public sector, in Finland and elsewhere. Our analysis of Suomi.fi services shows that basic architecture of digital services uses platform technology, like X-gate data exchange technology. Similarly, tools and resources for service providers are available, making it easier to enter into service platforms. So far so good, but what is lacking is the application of artificial intelligence to big data available in huge registers and data collected in the public sector. It is known that artificial intelligence has been used in military and security affairs (cyber wars, Owen 2015) as well as in management of energy production and consuming (smart grids). However

[5] About the health and social services reform, see http://alueuudistus.fi/en, accessed October 20, 2017.

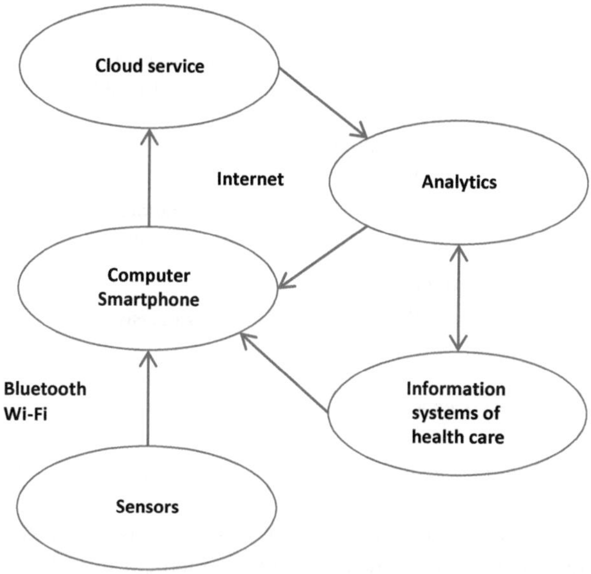

Fig. 2 The architecture of client-centred digital health care

applications, for example, in health care, are still in their infancy, although potential benefits are considerable.

We take health care as a special case for potential applications of artificial intelligence and big data analysis. As a recent report shows, the architecture of a digital health care is quite well defined, and all components needed to implement it are available (Hautamäki 2017). These components include *sensors* that sense changes in patient's condition and send signals to *computers* or *smartphones*. These computers send data to a *cloud service*, in which artificial intelligence system (analytics) makes analysis of data based on big databases. After this analysis, the results are integrated into an information system for health-care authorities. In addition, the results are transmitted back to a patient in a suitable form (Fig. 2).

This architecture is implemented partly in so-called self-care systems, which help people manage their wellbeing using many kinds of measurement instruments. Especially in athletics and exercise training, self-monitoring devices are in extensive use. These devices are provided by many brands, like Apple (USA), Polar Electro (Finland), Samsung (South Korea) and Suunto (Finland). We can divide the use of self-monitoring devices into two different groups (Hautamäki 2017). One group consists of voluntary use of devices for wellbeing and illness prevention. The other group consists of official medical uses controlled and funded by health-care institutions.

In group 1 people pay for these self-care devices by themselves and use them to follow their activity, training, dream intensity, heartbeat count, walking activity, etc. The quality control of these devices is not rigorous, and appropriate use of them is

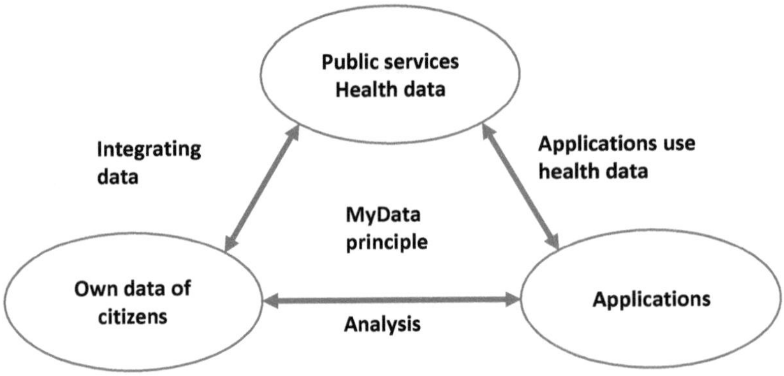

Fig. 3 The open architecture of digital health care

the sole responsibility of the user. Self-care is now a big trend and wearable devices are selling well.[6]

In group 2 the control of devices is extensive and rigorous: they must pass several tests before they are accepted for medical use. The health-care system has been using these devices as an integrated part of their system. For example, new devices like glucose metres and control programmes have been developed to help diabetes patients control their blood sugar levels.

Our analysis of self-care services is that while the market of the devices of group 1 is vast and growing, a greater benefit will be attained when these devices are integrated into the entire health-care system (Hautamäki 2017). Then the data produced by sensors and smartphones could be evaluated and analysed through big data and high-level analytics. This would allow using artificial intelligence in analysis and help develop new care for even rare diseases.

The Kanta service in Finland will provide a platform to implement the architecture of digital health care. There are two important elements being developed now. On the one hand, data produced by citizens, say by self-monitoring, must be embedded into official patient data to form a unified database. On the other hand, the database must be open to different applications so that a citizen can use the applications she likes to analyse the data. We call this system the open architecture of digital health care (Fig. 3). This system has not yet been implemented, but its principles are accepted by authorities.

[6] http://www.fiercehealthcare.com/it/self-care-med-devices-market-to-hit-16-8-billion-by-2019, accessed April 2 2017.

MyData

An important element of platform architecture adopted in Finland is MyData principle. MyData is a Finnish initiative presented in 2009 in order to develop rules of using personal data in government and business. The core of MyData allows individuals to control their own data. "This simplifies data flow and opens new opportunities for businesses to develop innovative personal data based services while preserving privacy" (https://mydatafi.wordpress.com/).

Alex Pentland has developed a similar approach to personal data calling it a "New Deal on Data" (Pentland 2014). The idea is to give individual citizens the rights to control their own personal data: citizens own their own data. Pentland explains the content of the "New Deal on Data" based on three principles:

1. You have the right to possess data about yourself.
2. You have the right to full control over the use of your data.
3. You have the right to dispose of or distribute your data.

MyData principle and a new deal on data are the precondition of successful development of digital services in the public sector as well as in business. There is also a need to build and enforce trust in these new digital platforms. Especially, if public services are adopting the open architecture described above, the legitimation of the system is a critical issue. Still we think that the MyData principle is easier to accept in the public sector than in business, because the public sector is under strong political control and all systems are transparent, in principle. But private business companies that own platforms have free access to all data produced by users. Global platforms, such as Google, Facebook and Twitter, benefit from the data produced tacitly by users of their services. Users do not know how and for what purposes their "own data" is used in business. These platforms apply artificial intelligence and sophisticated algorithms to analyse data and conduct business based on the results (Pentland 2014).

Another important aspect of the MyData principle is data security. All sharing of data and opening it to privately-owned applications involves a certain risk. In Finland, the national architecture of digital services contains many features necessary for security, like e-identification and e-authorisations. In communication between end users and service providers, SSL encryption is used. In the future, blockchain technology might be a useful tool for data security in public services like health care.

Pentland (2014) proposed devising "trust networks" for data sharing, involving "a combination of a computer network that keeps track of user permissions for each piece of personal data, and a legal contract that specifies both what can and can't be done with the data, and what happens if there is a violation of the permissions" (Pentland 2014, 182). In this system all personal data have attached labels specifying what one can do with the data. Trust networks are used in the interbank money transfer system, but they have not been available for general use.

Conclusion

The platform economy is dramatically transforming the business environment. The more services are digitalised, the more they will be produced and distributed on digital platforms. The owners of platforms are in a privileged position to earn substantial profits (compare Apple Store and iTunes). Also the data collected from transactions is extremely valuable (Google, Facebook). Notably, the application of artificial intelligence to big data will lead to many innovations unattainable so far (IBM's Watson system). Cloud computing allows service providers the freedom to concentrate on their core business. In summary, the development of new digital technology has created a rich toolbox to develop new kinds of services.

The platform economy is becoming an important tool for transforming the public sector and government. All new technologies are available and mostly well developed and tested in business, but the application of a new technology in the public sector is not a direct or certain process. There are special requirements concerning security, accessibility, affordability and availability. All people must be in an equal position regarding public services. The goal of "going digital" is not enough; people must also have the skills, capacity and tools to fully utilise digital services.

The most promising application of the platform economy is a unified, single platform for all digital services provided by the government. A Finnish public portal, Suomi.fi, is an example of such an approach. Although the palette of services is wide, the number of potential users is relatively small. The "network effect" has not yet been fully realised; thus the digitalisation of services has not yet resulted in savings, which was anticipated. The general impression about the development of digital public services is that they are mostly created from the system's viewpoint. Therefore the services are not easy to use and not as attractive as they should to be in order to reach the critical mass of citizens.

The experimental approach and the adoption of design thinking (Brown 2009) are right steps towards public digital platforms, which are citizen-centred and largely accepted as a viable alternative to traditional services. In the experimental approach, service design is becoming one of the major tools to develop public services (Annala et al. 2015). Solutions are increasingly produced by co-creation of authorities, citizens and companies, via the so-called Public People Private Partnership. These steps are even more important if the aim is to develop and use public platforms owned by public organisations. The technology and systems needed are usually provided by private companies, but the design of systems can be in public control. We argue that algorithms behind the public platform economy should not be business secrets – this way the anticipated algorithmic and artificial intelligence revolution will not be out of control. Perhaps, we need a principle comparable to the MyData principle related to algorithms: Our algorithm demands that all algorithms used in public services be transparent and open to changes arising from experiences.

References

Annala, M., Kaskinen, T., Lee, S., Leppänen, J., Mattila, K., Neuvonen, A., Nuutinen, J., Saarikoski, E., & Tarvainen, A. (2015). *Design for government: Human-centric governance through experiments*. Publication as a part the Government's Analysis, Assessment and Research Activities 7/2015, Helsinki.

Behavioural Insights Team. (2017). Website accessed 14 Mar 2017. http://www.behaviouralinsights.co.uk/

Bell, E., & Owen, T. (2017). *The platform press: How Silicon Valley reengineered journalism*. New York: Tow Center for Digital Journalism, Columbia Journalism School.

Breckon, J. (2015). *Better Public Services through experimental government*. Alliance, Nesta [online] available from http://www.alliance4usefulevidence.org/assets/Final.pdf. Accessed 23 Mar 2017.

Brown, T. (2009). *Change by design. How design thinking transforms organisations and inspires innovation*. New York: HarperCollins Publishers.

Chesbrough, H. (2011). *Open services innovation, rethinking your business to grow and compete in a new era*. San Francisco: Jossey-Bass.

Daglio, M., Gerson, D., & Kitchen, H. (2014, November 12–13). *Building organisational capacity for public sector innovation*. Background paper prepared for the OECD Conference "Innovating the Public Sector: From Ideas to Impact", Paris.

Eisenmann, T., Parker, G., & Van Alstyne, M. (2006). Strategies for two-sided markets. *Harvard Business Review, 84*(10), 92–201.

Experimental Finland. (2017). Website accessed 23 Feb 2017. http://kokeilevasuomi.fi/en

Gawer, A. (2014). Bridging differing perspectives on technological platforms: Toward an integrative framework. *Research Policy, 43*(7), 1239–1249.

Gore, E. M. (2014). Nonprofits should lead the sharing economy. *Standford Social Innovation Review*. https://ssir.org/articles/entry/nonprofits_should_lead_the_sharing_economy. Accessed 28 Apr 2017.

Hautamäki, A. (2017). *Terveyden omaehtoinen edistäminen – kehityspolku* [Self-care – The development path]. Tekesin katsauksia 326/2017.

Hautamäki, A., & Oksanen, K. (2015). Systemic development of service innovation. In R. Agarwal, W. Selen, G. Roos, & R. Green (Eds.), *The handbook of service innovation* (pp. 349–371). London: Springer.

Ikävalko, P. (2017, May 6). Omat terveystiedot näkee Omakannasta [Own health data is seen from Omakanta]. *Helsingin Sanomat Newspaper*.

Kenney, M., & Zysman, J. (2016). The rise of the platform economy. *Issues in Science and Technology 32*(3), Spring, 61–69.

Kushida, K. E., Murray, J., & Zysman, J. (2014). Cloud computing: From scarcity to abundance. *Journal of Industry, Competition and Trade*. https://doi.org/10.1007/s10842-014-0188-y.

Lourenço, J. S., Ciriolo, E., Almeida, R. S., & Troussard, X. (2016). *Behavioural insights applied to policy: European Report 2016*. EUR 27726 EN.

OECD. (2015). *The innovation imperative in the public sector: Setting an agenda for action*. Paris: OECD Publishing. https://doi.org/10.1787/9789264236561-en

OECD. (2017, February 12–14). *Embracing innovation in government. Global trends*. Dubai: World Government Summit in collaboration with the OECD. http://www.oecd.org/innovation/innovative-government/embracing-innovation-in-government-global-trends.htm

Owen, T. (2015). *Disruptive power, the crisis of the state in the digital age*. Oxford: Oxford Univeristy Press.

Pentland, A. (2014). *Social physics, how good ideas spread – The lessons from a new science*. New York: The Penguin Press.

PricewaterhouseCoopers. (2007, January). *The road ahead for public service delivery: Delivering on the customer promise*. Luxembourg: The Public Sector Research Centre, PwC.

Sandararajan, A. (2013). From Zipcar to the sharing economy. *Harvard Business Review*. http://oz.stern.nyu.edu/SharingEconomy2013HBR.pdf. Accessed 28 Apr 2017.
Smedlund, A. (2016). *Digital health platform complementor motives and effectual reasoning*. In 49th Hawaii International Conference on System Sciences (HICSS), pp. 1614–1623.
The Innovation Policy Platform (IPP). (2017). *IPP platform by the OECD and the World Bank*. https://www.innovationpolicyplatform.org/content/public-sector-innovation
Tiwana, A. (2014). *Platform ecosystems: Aligning architecture, governance, and strategy*. Waltham: Elsevier.
Vazquez Sampere, J. P. (2016, April 8). Why platform disruption is so much bigger than product disruption. *The Harvard Business Review*. https://hbr.org/2016/04/why-platform-disruption-is-so-much-bigger-than-product-disruption
Viskari, J. (2017, March 20). Sähköiset palvelut on sovitettava yhteen, [Electrical services have to fit together]. *Helsingin Sanomat Newspaper*.
Zysman, J. (2006). The 4th service transformation: The algorithmic revolution. *Communications of the Association for Computing Machinery: Special Issue on Services Sciences, 49*(7):48.
Zysman, J. (2014). *Escape from the commodity trap: Will the production transformation sustain productivity, growth and jobs? A manuscipt*. Berkeley: Brie.

Websites

http://kokeilevasuomi.fi/en
http://vrk.fi/en/ (the Population Register Centre of Finland)
http://www.suomi.fi/suomifi/english/index.html
https://mydatafi.wordpress.com
www.kanta.fi/en/
www.vm.fi/en/ (the Ministry of Finance in Finland)

A Conceptual Platform for Understanding and Managing Complex Service Enterprises: Case Studies in the Transformation of Healthcare Delivery

William B. Rouse, Michael J. Pennock, Zhongyuan Yu, and Kara M. Pepe

Abstract The nature of service delivery enterprises is explored in the context of transforming such enterprises to achieve substantially improved effectiveness and efficiency in delivering service outcomes. The impacts of the architectures of service enterprises are considered in terms of multilevel enterprise models. A methodology for developing such models is presented. Computational versions of such models are discussed. Five case studies are summarized that illustrate the application of this conceptual platform to several aspects of health and healthcare.

Keywords Enterprise transformation · Multi-level modeling · Computational modeling · Healthcare delivery

Introduction

Service enterprises in labor-intensive industries such as education, health, and government face a variety of challenges. In these industries, technological innovation has tended to increase costs as consumers' demands for technologically enabled services increase without the associated decreases in labor costs usually experienced by other industries. This has been termed "cost disease" (Baumol and Bowen 1966; Baumol 1967).

This phenomenon makes it difficult to pursue goals such as universal education and health. Significant improvements of efficiency are needed for such coveted goals to be affordable. However, service enterprises are both enabled and constrained by the architectures they operate within. As Deming (1986) convincingly argues, service workers' performance is not only affected by their abilities and moti-

W. B. Rouse (✉) · M. J. Pennock · Z. Yu · K. M. Pepe
Stevens Institute of Technology, Hoboken, NJ, USA
e-mail: wrouse@stevens.edu

© Springer Nature Singapore Pte Ltd. 2018

A. Smedlund et al. (eds.), *Collaborative Value Co-creation in the Platform Economy*, Translational Systems Sciences 11,
https://doi.org/10.1007/978-981-10-8956-5_6

vation. The organizational systems they operate within both enable and constrain them. This is true of human performance at all levels of the enterprise, from front-line service operations to the executive suite.

Before considering how the architecture of an enterprise affects service enterprises, it is useful to discuss the pressures they encounter when trying to address fundamental needs to change. Once this context is elaborated, we will return to architecture and modeling concerns. An overall approach to multilevel modeling is presented. Computational solutions of these models are next discussed. Five case studies of healthcare delivery are presented, followed by consideration of the general implications of this computational approach for transforming complex service enterprises.

It is important to preface the upcoming discussions with a comment on platforms. This chapter does not discuss an explicit technology platform like a portable device or a vehicle. The platform in this chapter is conceptual in that it provides a way of thinking about complex service systems that yields a clear path to instantiations of computational models that can support decision-makers as they entertain substantial changes of these systems.

Enterprise Transformation

Our earlier studies (Rouse 2005, 2006) have led us to formulate a qualitative theory, "Enterprise transformation is driven by experienced and/or anticipated value deficiencies that result in significantly redesigned and/or new work processes as determined by management's decision making abilities, limitations, and inclinations, all in the context of the social networks of management in particular and the enterprise in general."

There is a wide range of ways to pursue transformation. Figure 1 summarizes conclusions drawn from a large number of case studies. The ends of transformation can range from greater cost efficiencies to enhanced market perceptions, to new product and service offerings, and to fundamental changes of markets. The means can range from upgrading people's skills to redesigning business practices, to significant infusions of technology, and to fundamental changes of strategy. The scope of transformation can range from work activities to business functions, to overall organizations, and to the enterprise as a whole.

The framework in Fig. 1 has provided a useful categorization of a broad range of case studies of enterprise transformation. Considering transformation of markets, Amazon leveraged IT to redefine book buying, while Wal-Mart leveraged IT to redefine the retail industry. In these two instances at least, it can be argued that Amazon and Wal-Mart just grew; they did not transform. Nevertheless, their markets were transformed.

Illustrations of transformation of offerings include UPS moving from being a package delivery company to a global supply chain management provider, IBM's transition from manufacturing to services, Motorola moving from battery elimina-

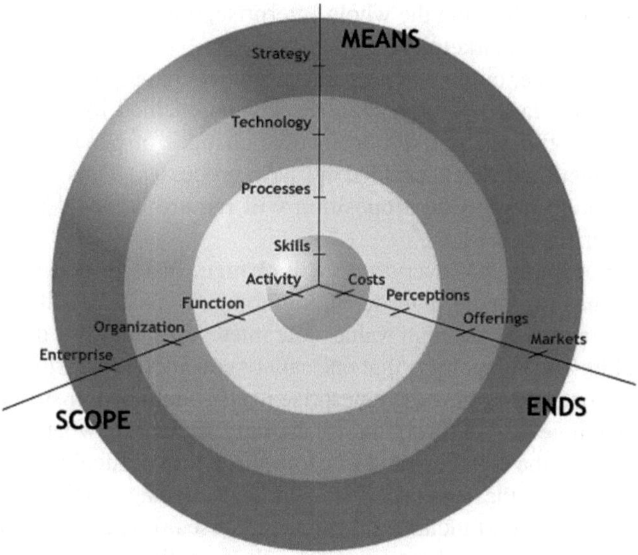

Fig. 1 Transformation framework

tors to radios to cell phones, and CNN redefining news delivery. Examples of transformation of perceptions include Dell repositioning computer buying, Starbucks repositioning coffee purchases, and Victoria's Secret repositioning lingerie buying. The many instances of transforming business operations include Lockheed Martin merging three aircraft companies, Newell Rubbermaid resuscitating numerous home products companies, and Interface adopting green business practices.

The costs and risks of transformation increase as the endeavor moves farther from the center in Fig. 1. Initiatives focused on the center will typically involve well-known and mature methods and tools from industrial engineering and operations management. In contrast, initiatives toward the perimeter will often require substantial changes of products, services, channels, etc., as well as associated large investments.

It is important to note that successful transformations in the outer band of Fig. 1 are likely to require significant investments in the inner bands also. In general, any level of transformation requires consideration of all subordinate levels. Thus, for example, successfully changing the market's perceptions of an enterprise's offerings is likely to also require enhanced operational excellence to underpin the new image being sought. As another illustration, significant changes of strategies often require new processes for decision-making, e.g., for R&D investments.

The transformation framework can be applied to thinking through a range of scenarios. The inner circle in Fig. 1 focuses on enterprise efficiency by, for example, focusing on particular activities, the skills needed for these activities, and the costs of these activities. In contrast, the outer circle of Fig. 1 might focus on totally new value propositions, addressing the whole enterprise, rethinking strategy, and fundamentally changing the marketplace.

Changes in the outer circle will very likely require changes in the adjacent circle. New offerings in a range of organizations will be enabled by new technologies. Success of these offerings is likely to involve changes of perceptions in the next circle at the functional level, enabled by new processes. Thus, we can see that embracing a totally new value proposition will require reconsideration of everything the enterprise does.

This does not imply that everything will change. Instead, it means that everything needs to be considered in terms of how things consistently fit together, function smoothly, and provide high value outcomes. This may be daunting, but is entirely feasible. The key point is that one cannot consider transforming the marketplace without considering how the enterprise itself should be transformed.

We hasten to note that, at this point, we are only addressing what is likely to have to change, not how the changes can be accomplished. In particular, we are not considering how to gain the support of stakeholders, manage their perceptions and expectations, and sustain fundamental change (Rouse 2001, 2006, 2007).

In this chapter, we argue that scenarios such as outlined above can be explored computationally (Rouse and Boff 2005). Thus, we are proposing that computational transformation should precede physical transformation. This enables exploration of a wide range of scenarios and, in particular, helps to get rid of bad ideas quickly. The good ideas that remain can be carefully refined for empirical investigation.

Multilevel Models

We need to computationally model the functioning of the complex enterprise of interest to enable decision-makers, as well as other significant stakeholders, to explore the possibilities and implications of transforming their enterprises in fundamental ways. The goal is to create organizational simulations that will serve as "policy flight simulators" for interactive exploration by teams of often disparate stakeholders who have inherent conflicts, but need and desire an agreed-upon way forward (Rouse and Boff 2005; Rouse 2014).

Consider the architecture of complex enterprises, defined broadly, shown in Fig. 2 (Rouse 2009; Rouse and Cortese 2010; Grossman et al. 2011). The efficiencies that can be gained at the lowest level (work practices) are limited by nature of the next level (delivery operations). Work can only be accomplished within the capacities provided by available processes. Further, delivery organized around processes tends to result in much more efficient work practices than for functionally organized business operations.

However, the efficiencies that can be gained from improved operations are limited by the nature of the level above, i.e., system structure. Functional operations are

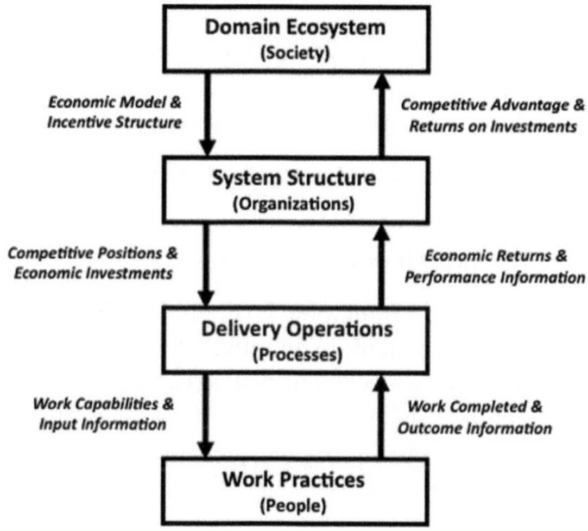

Fig. 2 Architecture of complex enterprises

often driven by organizations structured around these functions, e.g., marketing, engineering, manufacturing, and service. Each of these organizations may be a different business with independent economic objectives. This may significantly hinder process-oriented thinking.

And, of course, potential efficiencies in system structure are limited by the ecosystem in which these organizations operate. Market maturity, economic conditions, and government regulations will affect the capacities (processes) that businesses (organizations) are willing to invest in to enable work practices (people), whether these people be employees, customers, or constituencies in general. Economic considerations play a major role at this level (Rouse 2010a, b).

These organizational realities have long been recognized by researchers in sociotechnical systems (Emery and Trist 1973), as well as work design and system ergonomics (Hendrick and Kleiner 2001). Policy flight simulators enable computational explorations of these realities, especially by stakeholders without deep disciplinary expertise in these phenomena. Empowering decision-makers to fly the future before they write the check can dramatically increase confidence and commitment to courses of action.

Modeling Methodology

We have developed (Rouse 2015) and evaluated (Pennock et al. 2017) a methodology for developing multilevel models. The ten-step enterprise modeling methodology was developed to allow decision-makers and policy makers to:

- Identify the key drivers of system behavior and resulting outcomes.

- Perform "what-if" analyses.
- Evaluate the efficacy of policy options to alter system behavior and outcomes.
- "Test drive" the future.
- Allow key stakeholders to experience the behavior of the "to-be" system.

The full methodology is documented in Rouse (2015). We will briefly summarize the ten steps below:

Step 1: Decide on the Central Questions of Interest

The history of modeling and simulation is littered with failures of attempts to develop models without clear intentions in mind. Models provide means to answer questions. Efforts to model socio-technical systems are often motivated by decision-makers' questions about the feasibility and efficacy of decisions on policy, strategy, operations, etc. The first step is to discuss the questions of interest with the decision-maker(s), define what they need to know to feel that the questions are answered, and agree on key variables of interest.

Step 2: Define Key Phenomena Underlying These Questions

The next step involves defining the key phenomena that underlie the variables associated with the questions of interest. Phenomena can range from physical, behavioral, or organizational to economic, social, or political. Broad classes of phenomena across these domains include continuous and discrete flows, manual and automatic control, resource allocation, and individual and collective choice. Mature domains often have developed standard descriptions of relevant phenomena.

Step 3: Develop One or More Visualizations of Relationships Among Phenomena

Phenomena can often be described in terms of inputs, processes, and outputs. Often the inputs of one phenomenon are the outputs of other phenomena. Common variables among phenomena provide a basis for visualization of the set of key phenomena. Common visualizations methods include block diagrams, IDEF, influence diagrams, and systemigrams.

Step 4: Determine Key Tradeoffs that Appear to Warrant Deeper Exploration

The visualizations resulting from Step 3 often provide the basis for in-depth discussions and debates among members of the modeling team as well as the sponsors of the effort, which hopefully includes the decision-makers who intend to use the results of the modeling effort to inform their decisions. Lines of reasoning, perhaps only qualitative, are often verbalized that provides the means for immediate resolution of some issues, as well as dismissal of some issues that no longer seem to matter. New issues may, of course, also arise.

Step 5: Identify Alternative Representations of These Phenomena

Computational representations are needed for those phenomena that will be explored in more depth. These representations include equations, curves, surfaces, process models, agent models, etc. – in general, instantiations of standard representations.

Boundary conditions can affect choices of representations. This requires deciding on fixed and variable boundary conditions such as GDP growth, inflation, carbon emissions, etc. Fixed conditions can be embedded in representations, while variable conditions require controls such as slider bars to accommodate variations – see Step 9.

Step 6: Assess the Ability to Connect Alternative Representations

Representations of phenomena associated with tradeoffs to be addressed in more depth usually require inputs from other representations and produce outputs required by other representations. Representations may differ in terms of dichotomies such as linear vs. nonlinear, static vs. dynamic, deterministic vs. stochastic, continuous vs. discrete, and so on. They may also differ in terms of basic assumptions, e.g., Markov vs. non-Markovian processes. This step involves determining what can be meaningfully connected together.

Step 7: Determine a Consistent Set of Assumptions

The set of assumptions associated with the representations that are to be computationally connected need to be consistent for the results of these computations to be meaningful. At the very least, this involves synchronizing time across representations, standardizing variable definitions and units of measures, and agreeing on a common coordinate system or appropriate transformations among differing coordinate systems. It also involves dealing consistently with continuity, conservation, and independence assumptions.

Step 8: Identify Datasets to Support Parameterization

The set of representations chosen and refined in Steps 5–7 will have parameters such as transition probabilities, time constants, and decay rates that have to be estimated using data from the domain(s) in which the questions of interest are to be addressed. Data sources need to be identified and conditions under which these data were collected determined. Estimation methods need to be chosen, and in some cases developed, to provide unbiased estimates of model parameters.

Step 9: Program and Verify Computational Instantiations

To the extent possible, this step is best accomplished with commercially available software tools. The prototyping and debugging capabilities of such tools are often well worth the price. A variant of this proposal is to use commercial tools to prototype and refine the overall model. Once the design of the model is fixed, one can then develop custom software for production runs. The versions in the commercial tools can then be used to verify the custom code. This step also involves instantiating interactive visualizations with graphs, charts, sliders, radio buttons, etc.

Step 10: Validate Model Predictions, at Least against Baseline Data

The last step involves validating the resulting model. This can be difficult when the model has been designed to explore policies, strategies, etc. for which there inherently is no empirical data. A weak form of validation is possible by using the model to predict current performance with the "as-is" policies, strategies, etc. In general, models used to explore "what-if" possibilities are best employed to gain insights that can be used to frame propositions for subsequent empirical study.

Summary

The logic of the ten-step methodology can be summarized as follows, with emphasis on Steps 1–7:

- Define the question(s) of interest.
- Identify relevant phenomena.
- Visually compose phenomena.
- Identify useful representations.
- Computationally compose representations.

Note that this logic places great emphasis on problem framing and formulation. Deep computation is preserved for visually identified critical tradeoffs rather than the whole problem formulation. Steps 8–10 of the methodology are common to many methodologies.

Not all problems require full use of this ten-step methodology. Application to any given problem may result in one, multiple, or no computational models. Often visual portrayals of phenomena and relationships are sufficient to provide the insights of interest. As just noted, such views are also valuable for determining which aspects of the problem should be explored more deeply. This is often where the full range of problem stakeholders is highly involved.

Stakeholder Involvement

The value of the methodology presented in this section tends to be directly proportional to the involvement of key stakeholders in the process, which can include both decision-makers and subject matter experts. They need to be intimately involved in Step 1, deciding on the central questions of interest. During Steps 3 and 4, their insights and needs should determine which tradeoffs can be resolved solely using the visualizations and which warrant deeper computational exploration. In Step 10, validation, their perceptions, and insights are obviously central.

Ideally, a few key stakeholders should be members of the team developing the models and visualizations. They need to perform frequent "sanity tests" of the emerging policy flight simulator. They need to take the controls and run the simulator through its paces. In particular, they need to find the conditions where the simulator yields results that are obviously wrong and then work with the team to diagnose the sources of these problems and fix them.

Having worked with thousands of decision-makers in well over 100 companies and agencies, I have never encountered a decision-maker who said something like, "Well, the models' predictions make no intuitive sense, but the computer must be right, so I will do what it suggests." Instead, there were two questions that were frequently asked (Rouse 2001).

First, "How wrong can I be and still have this decision make sense?" This question reflects underlying uncertainties about stakeholders' and their intentions, attributes, and relative importance and alternatives and their attributes. They knew some assumptions would inevitably turn out to be wrong. They wanted to know how sensitive the "goodness" of the decisions being entertained was to the various assumptions.

Second, "How bad can things get and still have this decision make sense?" For this question, the underlying uncertainties concern consequences and their implications, stakeholders' reactions to consequences, and abilities to influence consequences. They knew there would be unexpected consequences of the decisions about to be made. They wanted to understand what kinds of consequences could undermine the goodness of these decisions.

In general, decision-makers were skeptical of model-based predictions. They did not expect that the future could be predicted with any degree of accuracy. They did, however, feel comfortable with the idea that model-based predictions could provide insights into what *might* happen and the conditions under which these outcomes *might* be precipitated.

Computational Solutions

We have employed the above methodology to develop several multilevel models of complex socio-technical systems over the past few years (Rouse 2015, 2016). Multilevel models are used to develop conceptual models that involve multiple levels of abstraction to represent the phenomena of interest. The term "level" concerns differing abstractions, rather than any inherently spatial differentiations.

The computational models that result involve connecting models of the phenomena of interest. Computationally, the notion of levels tends to disappear. Connecting models can pose "composition" difficulties. Issues range from consistency of variable definitions, units of measure, and coordinate systems to much more difficult problems ranging from conflicting assumptions regarding independence, conservation, and continuity to the different models having entangled states.

There appears to be no general approach to resolving composition difficulties. For each multilevel modeling instance, one has to identify these issues and determine how to correct inconsistencies or at least work around them to preserve overall model validity. This tends to be easier when one is working directly with the mathematics of each model, rather than legacy software code. Composition of software codes presents fundamental problems far beyond getting the resulting composition to compute (Tolk 2014).

Composition of sets of mathematical equations enables easier inspection of the component equations. Figure 3 depicts an example approach for composing a mul-

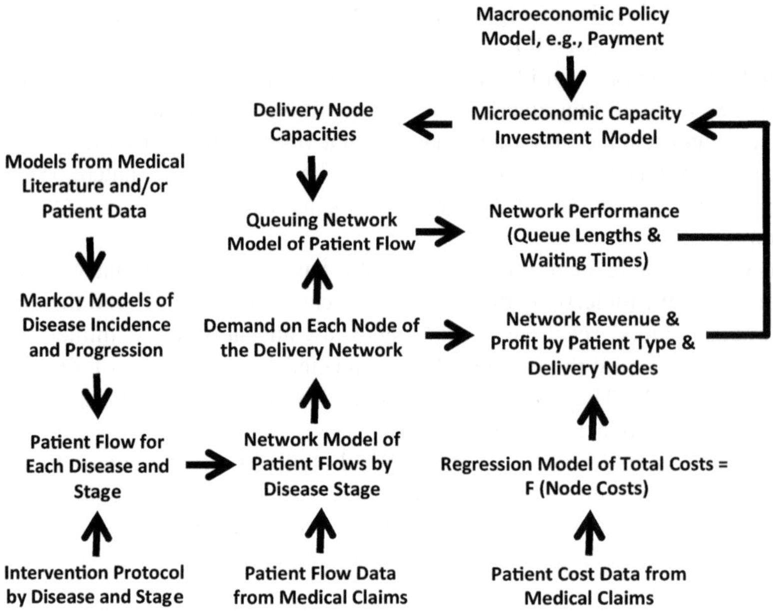

Fig. 3 An approach modeling healthcare delivery

tilevel model for healthcare delivery. This example is included here to illustrate a few, but not all, composition issues in a bit more detail.

We need to predict patient flows for each disease and stage. This enables predicting patient flows which feed a network model of patient flows to service nodes, informed by intervention protocols. The network model is used to predict demands on each service node.

We next need to predict the performance of this network. Given arrival rates, service rates, numbers of servers, and queue disciplines, one can compute mean time spent in the system, mean number of entities in the system, mean time spent queuing, and mean number of entities in queues (White et al. 2012). These predictions can inform decisions regarding investments in service delivery capacities relative to the revenue and profit potential of these services.

It is usually fairly straightforward to assure the consistency of the probabilistic assumptions underlying the various models depicted in Fig. 3. Thus, at least one composition issue is handled. However, we still need to be concerned with the validity of the overall set of assumptions. For example, how can the assumption of random flows be justified in a system where appointments are scheduled? It has been found that arrivals can be modeled as random because of all the variability that creeps in along the way. Of course, if this assumption cannot be justified, the queuing network model would have to be adjusted.

The central tradeoff in Fig. 3 involves the lower left versus upper right – designing intervention protocols by disease and stage versus macroeconomic policies on payments. Payment models have an enormous impact on investments in capacities, resulting network performance, and provider revenues and profits. For the provider, this tradeoff drives intervention protocols and investment strategies.

The approach embodied in Fig. 3 allows a completely analytic solution for expected values. Simulation is required if variance predictions are also needed. Of course, the analytic and simulation predictions of means should be identical. One can vary the probabilistic assumptions in the simulation and determine where variations of assumptions make a difference. This provides a much deeper understanding of both models.

As noted in the earlier discussion of Step 9, Program and Verify Computational Instantiations, we often construct early prototypes in Microsoft Excel and full prototypes in AnyLogic. Once the full model is developed and tested, we sometimes reprogram it in Python or another language for hosting on the web. Comparisons of its outputs to the programs in the commercial tools can provide tests of the new code. The extent to which this process is followed depends on the deliverables required for any particular project.

All of our models are packaged and characterized as policy flight simulators (Rouse 2014, 2015; Rouse et al. 2016), with the aforementioned tag line, "Drive the future before you write the check." This approach, combined with stakeholders taking the controls and exploring alternative futures for their enterprise, greatly helps to overcome nontechnical stakeholders' difficulties of understanding and feeling confident in "black box" models and simulations (Yu et al. 2016).

Case Studies

Modeling and simulation have been extensively used to understand and improve healthcare delivery processes (Rouse and Cortese 2010; Rouse and Serban 2014). Many applications have been at the hospital level, focused on scheduling, staffing, patient flow, and, in general, efficient and effective use of various capacities. The five case studies summarized in this section were pursued over 5 years and contributed to the genesis of Figs. 2 and 3. The notion emerged of a computational platform for addressing a range of enterprise transformation endeavors, including healthcare, of course, but also higher education (Rouse 2016) and acquisition and operation of defense platforms (Pennock et al. 2017).

Emory: Prevention of DM and CHD

This case study addressed the employee prevention and wellness program of Emory University (Park et al. 2012). The application of the multilevel model focused on the roughly 700 people in this cohort and their risks of diabetes mellitus (DM) and coronary heart disease (CHD). Each person's risk of each disease was calculated using DM and CHD risk models from the medical literature, using initial individual assessments of blood pressure, fasting glucose level, etc. Subsequent assessment data were used to estimate annual risk changes as a function of initial risks of each disease.

The model of the healthcare delivery enterprise in Fig. 2 includes four levels – the *Ecosystem* level, the *Organization* level, the *Process* level, and the *People* level. Each level introduces a corresponding conceptual set of issues and decisions for both the payer and provider. In this case, the Human Resources Department of Emory University (HR) was the payer responsible for healthcare costs for university employees, while the Predictive Health Institute (PHI) was the provider focused on prevention and maintenance of employee health.

The *Ecosystem* level allows decision-makers to test different combinations of policies from the perspective of HR. For instance, this level determines the allocation of payment to PHI based on a hybrid capitated and pay-for-outcome formula. It also involves choices of parameters such as projected healthcare inflation rate, general economy inflation rate, and discount rate that affect the economic valuation of the prevention and wellness program. One of the primary concerns of HR is achieving a satisfactory ROI on any investments in prevention and wellness.

The concerns at the *Organization* level include the economic sustainability of PHI – their revenue must be equal to or greater than their costs. To achieve sustainability, PHI must appropriately design its operational processes and rules. Two issues are central. What risk levels should be used to stratify the participant population? What assessment and coaching processes should be employed for each strata of the population? Other *Organization* level considerations include the growth rate of the participant population, the age ranges targeted for growth, and the program duration before participants are moved to "maintenance."

The daily operations of PHI are represented on the *Process* level. Participants visit PHI every 6–12 months. Health partners employed by PHI perform assessments, work with participants to set health goals, and perform follow-up calls or emails to monitor participants and encourage them to follow their plan. This level is represented as a discrete-event queuing network.

The *People* level is the replication of the actual population of PHI participants. Over a 3-year period, roughly 700 participants joined this prevention and wellness program. Each of them had various assessment measurements recorded such as blood pressure, fasting glucose level, etc. Each participant was instantiated in the model as an agent. Based on the assessment measurements, the risk of developing DM or CHD was computed for each agent. Then, total healthcare costs were estimated for their remaining life based on their risk level for each disease. The reduced amount of aggregated total healthcare cost achieved by PHI is an *Ecosystem* level benefit to the HR organization.

The four-level model was implemented in AnyLogic Version 6.7. Runs of the multilevel simulation are set up using the dashboard in Fig. 4. Beyond the decision variables discussed above, decision-makers can decide what data source to employ to parameterize the models – either data from the American Diabetes Association (ADA) and American Heart Association (AHA) or data specific to Emory employ-

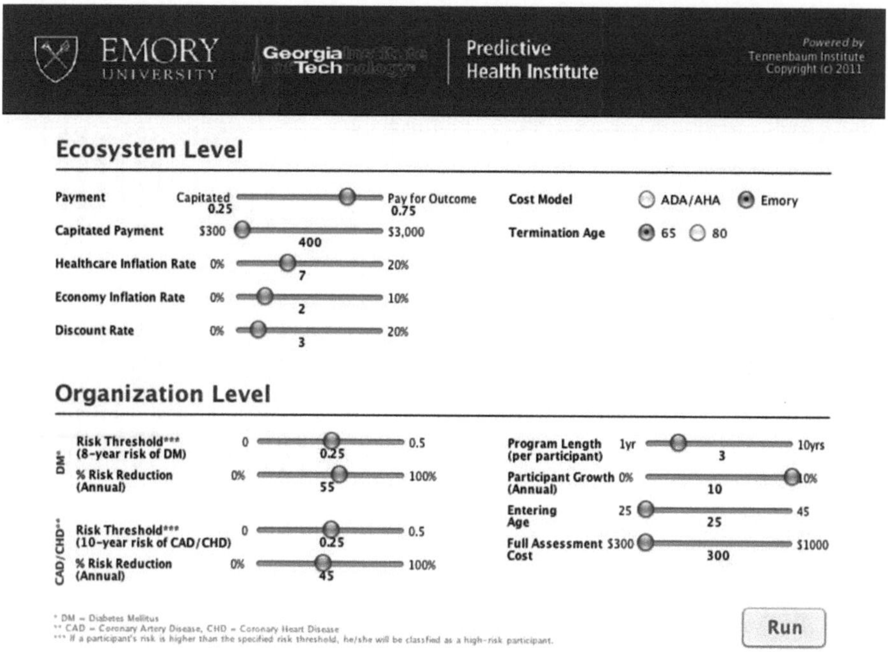

Fig. 4 Multilevel simulation dashboard

ees. Decision-makers can choose to only count savings until age 65 or also project postretirement savings.

The bottom half of the dashboard provides inputs from *Organization* level decision-makers, namely, PHI. Beyond the variables mentioned above, these decision-makers must choose how to stratify the participant population into low- and high-risk groups for each disease. Once they choose a level on the risk threshold slider, a set point appears on the percent risk reduction slider that represents what PHI is actually achieving based on analysis of their ongoing assessment data. Decision-makers can choose to operate at the set point by moving the slider to this point, or they can explore the consequences of larger or smaller risk reductions.

Figure 5 shows the *Ecosystem* and *Organization* levels of the model. The provider organization, PHI, decides how to stratify participant flows and seeks to have revenues equal or exceed costs. The payer organization, HR, sets the "rules of the game" as depicted on the dashboard in Fig. 4. HR's ROI from PHI's services is shown in net present values using the discount rate shown in Fig. 4.

The logic of the economic valuation provided by the model is as follows. PHI incurs costs of operating its processes to reduce the risks of DM and CHD for its population of participants. The resulting risk reductions delay the onset of these diseases for participants, often beyond their projected life span. This results in cost avoidance, both for treatment of these diseases and lost work productivity. This savings yields future cash flow to HR that enables them to provide revenue to

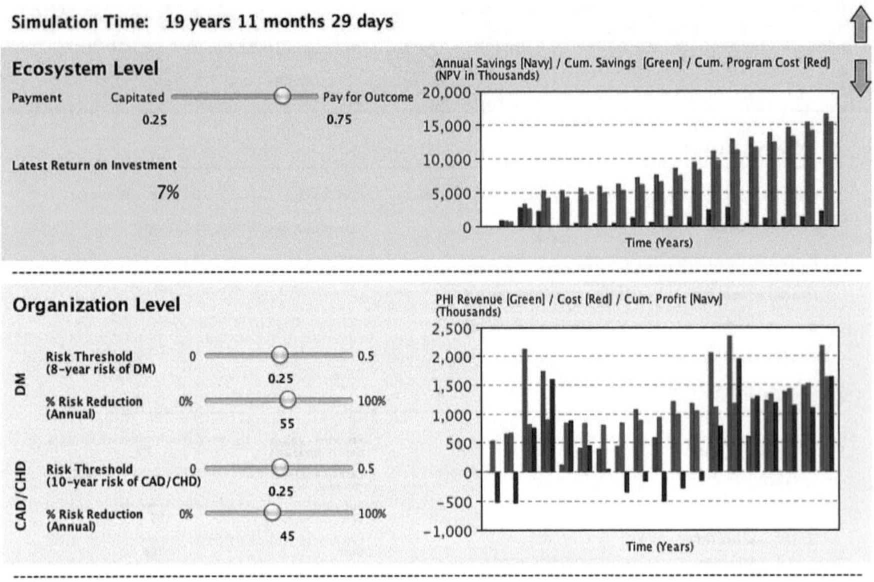

Fig. 5 Ecosystem and organization levels of model

PHI. However, as these savings will occur in the future and the investment must be made now, one needs to consider factors such as expected inflation. The result for PHI and HR consists of two time series each, one for costs and one for revenues. The difference between revenues and costs represents profit or loss. The net present value of this time series is then calculated using the discount rate from Fig. 4. The ROI shown in Fig. 5 is calculated from the latest (most recent year) ratio of savings to costs.

Of most interest are "economically attractive" configurations under which PHI is a sustainable organization and Emory HR also has a positive ROI. Several interesting results were found. All assume a participant risk stratification approach not actually used by PHI. Under the traditional approach, all participants receive the same full assessment and coaching program. The proposed risk stratification differentiates participants; only participants with greater than 25% risks of DM and/or CHD would receive the full assessment and coaching program.

When we compromise between the returns to HR and PHI, the aggregate returns to Emory are minimized. The best economic results are achieved when either PHI's profit is maximized or Emory HR's ROI is maximized. There are a variety of reasons why one might choose either extreme. One possibility was of great interest. HR could maximize its ROI while providing PHI a very lean budget. At the end of each year, HR could then provide PHI with a bonus for the actual savings experienced that year. This could be determined by comparing the projected costs for the people in the program to their actual costs of healthcare, absenteeism, and presenteeism. In this way, HR would be sharing actual savings rather than projected savings. The annual bonuses would free PHI of the fear of not being sustainable, although PHI would need to substantially reorganize its delivery system.

The implications of these results are clear. The financial objectives of HR and PHI – which are in conflict – should not be independently optimized; if either loses significantly, the system becomes dysfunctional. HR needs to adopt payment mechanisms under which PHI can redesign its delivery processes to achieve sustainability while also providing HR with an acceptable return on its investment in prevention and wellness.

In order for PHI to stay in business, it will have to change its business model, stratifying the population by risk levels and tailoring processes to each stratum. This could include an initial low-cost, streamlined assessment and subsequently PHI "Lite" for low-risk participants. PHI also needs to develop a low-cost "maintenance" process to sustain reduced risks once they have been achieved. These recommendations significantly influenced the subsequent redesign of PHI.

Vanderbilt: Chronic Disease Management for HTN, DM, and CHD

While the Emory case study focused on preventing DM and CHD, the Vanderbilt case study focused on managing these chronic diseases, as well as hypertension (HTN). We worked with the ***MyHealth Team*** at the Vanderbilt School of Medicine. This program brought together all the medical specialties needed to address patients' needs, combined with outreach capabilities to connect to patients in their home and take advantage of both phone and Internet.

There were several hundred patients enrolled in ***MyHealth Team*** at that time. In contrast to Emory where we had 3 years of clinical data for 700 participants, data was rather limited at Vanderbilt. Thus, estimating model parameters was challenging. This situation would, of course, get much better in time.

As is typical, the modeling team had many questions about processes and patient flow. Various diagrams were provided and many discussions ensued. It quickly became apparent that our questions were causing the iterative redesign of the delivery system. Consequently, what started as a model-building project became what is termed a "design-build" project. Put simply, the attempt to model the delivery processes was significantly affecting the design of the processes.

This can be viewed as an instance of model-based systems engineering. As such, the predictions of the models were less useful than the structured approach to thinking about processes, connections among processes, and implications for patient flows. The overall engagement was quite creative, but admittedly would have been even more productive had we realized at the outset the design-build nature of the endeavor.

Indiana: Chronic Disease Management for Alzheimer's Disease

This case study addressed care for patients with memory and emotional problems such as Alzheimer's disease and other related dementia. There is a substantial need to develop new scalable and sustainable brain care services to care for these patients. This care requires extensive psychosocial support, nursing care, and comprehensive patient-centered management, which strain the resources of clinicians, family caregivers, and community-based support structures. Indiana University developed such a health management program called the Aging Brain Care Medical Home (ABC) to provide the collaborative care model to 1500 older adults in central Indiana (LaMantia et al. 2014).

In order to scale up the ABC collaborative care model to more patients and other geographical areas, it was necessary to understand what factors affect the efficiency and effectiveness of its operations and outcomes. To this end, we developed a multilevel computer simulation model of the ABC program (Boustany et al. 2016). It was shown that scaling the program without modification would be infeasible. More broadly, the ABC simulation model served as a risk reduction decision support tool for healthcare delivery redesign, allowing early identification of operational and redesign pitfalls.

The ABC simulation model included elements from both agent-based and discrete-event modeling, incorporated in an overall multilevel model. The model was used to explore different strategies for scaling up the ABC program. Results showed that as population sizes increase, economies of scale are reached, and thus the contribution of fixed costs to the costs per member or per patient decreases. Another important finding that emerged from this study was that the impact of the ABC program on cost savings reaches a steady state after a period of several years, which is indicated by plateaued ROIs.

Penn: Transition Care for High-Risk Elderly Patients

The Transitional Care Model (TCM) is a proven care management approach that can contribute to a more person-centered, effective, and efficient response to the challenge of chronic illness. Despite TCM's proven value (Naylor 2012; Naylor et al. 2014), it has been challenging to convince decision-makers to implement this model. Success in TCM's spread has been achieved only slowly – one health system or community at a time. Among major barriers to widespread implementation are perceptions that the model has been demonstrated to work in randomized control and comparative effectiveness trials but not in the "real world", is too complex and costly, requires upfront investment which will largely benefit other providers downstream, or is not adaptable to local contextual issues.

While each of these misperceptions has been addressed through successful translation of the TCM in a number of health systems, traditional strategies (e.g., identifying local champions, multiple meetings with decision-makers) consume substantial time and are not as efficient as desired in promoting widespread scaling. Such challenges are not limited to the adoption of the TCM, and addressing them could have positive impacts on the widespread adoption of evidence care throughout the US healthcare system.

To that end, the specific goal of this case study was to determine whether the use of a policy flight simulator accelerates positive decisions to implement the TCM (Rouse et al. 2017a, b). As indicated earlier, policy flight simulators fuse aspects of scientific analysis, engineering, social science, and visualization to provide decision-makers with a more comprehensive understanding of the consequences of interventions than that provided by traditional mathematical and computational approaches.

To accomplish this goal, the team conducted two activities in an iterative, adaptive process. First, we elicited barriers and facilitators to adopting evidence-based, highly effective interventions from decision-makers representing providers, payers, and purchasers. Second, we developed and continuously refined the TCM policy flight simulator. In the process, several key insights emerged:

- The payment system is central.
- Beliefs about evidence vary; peers' actions are important.
- Research evidence is not sufficient.
- The offering must relate to "my population."

These insights caused us to realize that any investment decision of the magnitude of TCM would likely require the involvement of many stakeholders and organizations in a given healthcare system. Consequently, we elaborated our goal, namely, to determine whether the use of an innovative policy flight simulator would help healthcare decision-makers (providers, payers or purchasers) to make better-informed decisions regarding the adoption of TCM and increase their confidence in a decision to adopt TCM. Numerous demonstrations with senior executives from providers, payers, and purchasers resulted in continued elaboration and refinement of the TCM policy flight simulator, as well as many insights into how they could employ the simulator in their decision-making.

NYC Health Ecosystem: Mergers and Acquisitions

The Affordable Care Act (ACA) is causing a transformation of the healthcare industry in the USA. This industry involves complicated relationships among patients, physicians, hospitals, health plans, pharmaceutical companies, healthcare equipment companies, and government. Hospitals are uncertain about how they should best respond to threats and opportunities. This is particularly relevant for hospitals located in competitive metropolitan areas such as New York City, where a large number of hospitals are competing – many among the nation's best. Questions that arise in this uncertain environment include:

- What if we wait until the healthcare market stabilizes and only invest in operational efficiency?
- Should we merge with competing hospitals to increase negotiation power?
- Shall we only focus on acquiring physician practices in highly reimbursed diagnostic groups?

In this case study, we developed a data-rich agent-based simulation model (Yu et al. 2016) to study dynamic interactions among healthcare systems in the context of merger and acquisition (M&A) decision-making, where by "rich" we mean extensive rule sets and information sources, compared to traditional agent-based models. The proposed model includes agents' revenues and profitability (i.e., financial statements), operational performance and resource utilization, as well as a more detailed set of objectives and decision-making rules to address a variety of what-if scenarios.

We applied our modeling approach on M&A dynamics of hospitals in New York City, informed by in-depth data on 66 hospitals of the Hospital Referral Region in Bronx, Manhattan, and Eastern Long Island. The objective of the simulation model is to assist hospital administrators to assess the impact of implementing strategic acquisition decisions at the system level. This is accomplished by simulating strategies and interactions based on real historical hospital balance sheets and operational performance data.

The outcomes of the simulation include the number of hospitals remaining in the market and frequent M&A pairs of hospitals under various settings. By varying strategy inputs and relevant parameters, the simulation can be used to generate insights as to how these outcomes would change under different scenarios. The interactive visualizations complement the simulation model by allowing nontechnical users to interactively explore relevant information, to input parameter values for different scenarios, as well as to view and validate the results of the simulation model.

The merger and acquisition process was as follows. For each simulation period (year), M&A targets are identified for each acquiring hospital based on a set of six strategic drivers. Screening of candidates includes strategic and financial criteria. Acquisition offers were then estimated based on similar historical transactions. Through the process, teaching affiliation compatibilities between the two sides and regulations are checked for all deals. After a transaction is finalized, hospitals' attributes are updated in the model.

Interactive decision-makers vary input parameters to create and test different scenarios. In addition to the set of six strategic drivers, other input parameters include M&A transaction-related parameters and market predictions. Simulation outputs include (1) number and identities of hospitals that remain in the market, along with their financials and patient counts, and (2) a detailed list of M&A activities that happened throughout the simulation process.

Fig. 6 Merger and acquisition simulation in ***Immersion Lab***

We were able to determine the most frequently appearing M&A pairs and identify hospitals that have greater interest and capability in acquiring other hospitals. While there are other considerations relevant to hospital mergers and acquisitions that are not included in the simulation (e.g., payment structure), the simulation is adaptable enough to incorporate many of these for future studies.

The purpose of this simulation model is to serve as a means to facilitate strategic decision-making processes. To this end, we developed an interactive visualization environment, where market dynamics can be simulated and decision-makers can interact with different settings to address what-if scenarios. Users interact with several interactive visualizations concurrently in a large-scale interactive environment, where an array of seven touch-screen monitors provide a 8′-by-20′, 180 degree ***Immersion Lab*** as shown in Fig. 6. This environment provides nontechnically oriented stakeholders an immersive experience that greatly increases their comfort levels with data-driven decision-making.

The results from the simulation model facilitate M&A decision-making, particularly in identifying desirable acquisition targets, aggressive and capable acquirers, and frequent acquirer-target pairs. The frequencies of prevalent pairs of acquirer and target appearing under different strategies in our simulation are of particular interest. The frequency level is a relative value in that it depends on number of strategies included and hospitals involved. A high frequency suggests a better fit and also repeated attraction.

Validation of agent-based simulations is challenging, especially for high-level strategic decision simulations. The overall model and set of visualizations was validated in two ways. First, from a technical perspective, we compared our simulation results with Capital IQ's hospital merger and acquisition transaction dataset.

Although there is limited number of cases under our regional constraint in the Capital IQ's database, the realized M&A transactions appear in our results.

Second is the feedback from users. There were many, roughly 30, demonstrations to hospital decision-makers and healthcare consultants as well as senior executives from insurance, government, foundations, etc. In total, perhaps 200 people participated in the demos, and many took the controls and tried various options. They made many suggestions, and the number of types of interactive visualizations iteratively increased.

The key value of the overall model and set of visualizations is, of course, the insights gained by the human users of this environment. For example, they may determine the conditions under which certain outcomes are likely. They can then monitor developments to see if such conditions are emerging. Thus, they know what *might* happen, even though they cannot be assured what will happen. The greatest insights are gained not only from simulation but also from interactive visualizations that enable massive data exploration, which moves from a "one-size-fits-all" static report to more adaptable and useful decision process.

Summary

The five case studies summarized in this section represent an evolution of thinking and capabilities for multilevel modeling. We have learned how to leverage multiple levels of abstraction, while also learning that our computational instantiations of these models need not mirror these conceptual levels. We have also achieved a few aspects of reuse that can greatly enhance the value of a platform.

These case studies have explored enterprise transformation in several of the rings in Fig. 1. Three of these studies (Emory, Vanderbilt, Indiana) addressed the transformation of care delivery processes. The TCM case study focused on adoption decisions, which relate more to strategy than operations. The New York City study addressed industry consolidation, in other words the changing marketplace.

We are currently exploring population health and learning health systems (Rouse et al. 2017a). There is growing recognition of the social determinants of the health of a population (Cooper 2016). Succinctly, health is affected by a myriad of health, education, and social services. For example, the quality of housing affects population health (Butler et al. 2017). Population health can be literally mapped in terms of social determinants (Sullivan 2016). Consequently, it has been argued, for example, that Medicare payments should be linked to social risk factors (Buntin and Ayanian 2017). These insights have led us to conclude that population health involves integration of health, education, and social services to keep a defined population healthy, to address health challenges holistically, and to assist with the realities of being mortal. Realizing this vision will require a transformation of the US delivery system.

Conclusions

While all of the case studies addressed healthcare, the platform discussed in this chapter has general implications for transformation of complex service enterprises. Computationally, it is easy to imagine versions of Figs. 2 and 3 for other service industries, for example, higher education (Rouse 2016). Patients can be replaced with students or, in general, customers.

More profound, we think, is the impact of interactive visualizations on stakeholders' interactions, conclusions, and confidence in their deliberations. More specifically:

- When users take the controls, vary assumptions and preferences, and experience the consequences, they gain understanding and confidence in models and simulations.
- Interactive visualizations should be co-designed with envisioned users, as well as other key stakeholders, to assure valid, acceptable, and viable problem formulations, solution concepts, and degrees of freedom.
- Given the "what-if" nature of many questions, prediction-based insights are the goal rather than point predictions of the future; many insights relate to intuitively reasonable ideas that yield unexpected and undesirable outcomes.

We hasten to note that these are observations from a range of experiences rather than scientific experimental findings, which only rarely we have been able to pursue (Rouse 1998, 2007). Nevertheless, these qualitative conclusions have been repeatedly manifested in our experiences in a wide range of industries. Put simply, key stakeholders appear to embrace model-based, or evidence-based, decision-making when they understand it and feel very much involved.

Keep in mind, however, the earlier discussion of decision-makers' skepticism regarding model-based predictions. The best decision-makers and sponsors are those that ask many questions, often requiring "peeling back the onion" to determine how assumptions, data, and component models propagated through the overall model to yield the predictions of interest. Such probing and exploration typically lead to better models and visualizations.

References

Baumol, W. J. (1967). Macroeconomics of unbalanced growth: The anatomy of urban crises. *American Economic Review, 57*, 415–426.

Baumol, W. J., & Bowen, W. B. (1966). *Performing arts: The economic dilemma*. New York: Twentieth Century Fund.

Boustany, K., Pennock, M. J., Bell, T., Boustani, M., & Rouse, W. B. (2016). *Leveraging computer simulation models in healthcare delivery redesign*. Indianapolis: Indiana University Health, Aging Brains Care Center.

Buntin, M. B., & Ayanian, J. Z. (2017). Social risk factors and equity in Medicare payment. *New England Journal of Medicine., 376*(6), 507–510.

Butler, S. M., Matthew, D. B., & Cabello, M. (2017). *Re-balancing medical and social spending to promote health: Increasing state flexibility to improve health through housing*. Washington, DC: Brookings Institution.

Cooper, R. (2016). *Poverty and the myths of health care reform*. Baltimore: Johns Hopkins University Press.

Deming, W. E. (1986). *Out of crisis*. Cambridge, MA: MIT Press.

Emery, F., & Trist, E. (1973). *Toward a social ecology*. London: Plenum Press.

Grossman, C., Goolsby, W. A., Olsen, L., & McGinnis, J. M. (2011). *Engineering the learning healthcare system*. Washington, DC: National Academy Press.

Hendrick, H. W., & Kleiner, B. M. (2001). *Macroergonomics: An introduction to work system design*. Santa Monica: Human Factors and Ergonomics Society.

LaMantia, M. A., Alder, C. A., Austrom, M. G., Cottingham, A. J., Litzelman, D. K., Boustany, K. C., & Boustani, M. A. (2014). The aging brain care medical home program: Accomplishments and lessons learned at one year. *Alzheimer's & Dementia: The Journal of the Alzheimer's Association, 10*(4), P209.

Naylor, M. D. (2012). Advancing high value transitional care: The central role of nursing and its leadership. *Nursing Administration Quarterly, 36*, 115–126.

Naylor, M. D., Hirschman, K. B., Hanlon, A. L., et al. (2014). Comparison of evidence-based interventions on outcomes of hospitalized, cognitively impaired older adults. *Journal of Comparative Effectiveness Research., 3*(3), 245–257.

Park, H., Clear, T., Rouse, W. B., Basole, R. C., Braunstein, M. L., Brigham, K. L., & Cunningham, L. (2012). Multi-level simulations of health delivery systems. A prospective tool for policy, strategy, planning and management. *Journal of Service Science, 4*(3), 253–268.

Pennock, M. J., Bodner, D. A., & Rouse, W. B. (2017). Lessons learned from evaluating an enterprise modeling methodology. *IEEE Systems Journal*, in press.

Rouse, W. B. (1998). Computer support of collaborative planning. *Journal of the American Society for Information Science, 49*(9), 832–839.

Rouse, W. B. (2001). *Essential challenges of strategic management*. New York: Wiley.

Rouse, W. B. (2005). A theory of enterprise transformation. *Journal of Systems Engineering, 8*(4), 279–295.

Rouse, W. B. (Ed.). (2006). *Enterprise transformation: Understanding and enabling fundamental change*. Hoboken: John Wiley.

Rouse, W. B. (2007). *People and organizations: Explorations of human-centered design*. New York: Wiley.

Rouse, W. B. (2009). Engineering perspectives on healthcare delivery: Can we afford technological innovation in healthcare? *Journal of Systems Research and Behavioral Science, 26*, 1–10.

Rouse, W. B. (2010a). Impacts of healthcare price controls: Potential unintended consequences of firms' responses to price policies. *IEEE Systems Journal, 4*(1), 34–38.

Rouse, W. B. (Ed.). (2010b). *The economics of human systems integration: Valuation of Investments in people's training and education, safety and health, and work productivity*. New York: Wiley.

Rouse, W. B. (2014). Human interaction with policy flight simulators. *Journal of Applied Ergonomics, 45*(1), 72–77.

Rouse, W. B. (2015). *Modeling and visualization of complex systems and enterprises: Explorations of physical, human, economic, and social phenomena*. Hoboken: Wiley.

Rouse, W. B. (2016). *Universities as complex enterprises: How academia works, why it works these ways, and where the university enterprise is headed*. Hoboken: Wiley.

Rouse, W. B., & Boff, K. R. (Eds.). (2005). *Organizational simulation: From modeling and simulation to games and entertainment*. New York: Wiley.

Rouse, W. B., & Cortese, D. A. (Eds.). (2010). *Engineering the system of healthcare delivery*. Amsterdam: IOS Press.

Rouse, W. B., & Serban, N. (2014). *Understanding and managing the complexity of healthcare*. Cambridge, MA: MIT Press.

Rouse, W. B., Pennock, M. J., Oghbaie, M., & Liu, C. (2016). Interactive visualizations for decision support: Application of Rasmussen's abstraction-aggregation hierarchy. *Journal of Applied Ergonomics*, *47*, 59, 541–553. in press.

Rouse, W. B., Johns, M. M. E., & Pepe, K. (2017a). Learning in the healthcare enterprise. *Journal of Learning Health Systems*, in press.

Rouse, W. B., Naylor, M., Pennock, M. J., Yu, Z., Pepe, K. M., Hirschman, K. B., & Pauly, M. V. (2017b). *The use of policy simulation to implement the transitional care model*. Hoboken: Center for Complex Systems and Enterprises, Stevens Institute of Technology.

Sullivan, K. R. (2016). *Visualizing the social determinants of health*. Hoboken: MS Project, School of Systems and Enterprises, Stevens Institute of Technology.

Tolk, A. (2014). *Ontology, epistemology, and teleology for modeling and simulation: Philosophical foundations for intelligent M&S applications*. Berlin: Springer.

White, J. A., Schmidt, W., & Bennett, G. K. (2012). *Analysis of queueing systems*. New York: Elsevier.

Yu, Z., Rouse, W. B., Serban, S., & Veral, E. (2016). A data-rich agent-based decision support model for hospital consolidation. *Journal of Enterprise Transformation, 6*(3/4), 136–161.

Understanding Platform Transformations Through Routine Interactions

Lauri Paavola

Abstract The purpose of this chapter is to discuss and consider routines that operate on interfaces enabling transformation of and within platforms. We view these transformations through specific routine interactions, which enable modules and platforms to either bring about transformations or to respond to them. We do this by introducing the concept of transformational routines and justify that it provides micro-level insight into different cause-and-effect relationships. Moreover, while traditional theories of platform entity transformations tend to focus on general evolutionary outlines and continuous processes, transformational routines provide temporally and spatially limited settings in which to observe their critical turning points. Finally, with the help of an illustration on a case study of Tesco, a UK grocery retailer, we argue that all these properties enable efficient collection of rich data with applications to both routine and module- and platform-level analyses.

Keywords Transformational routines · Platforms · Modules · Interfaces · Ecosystems

Introduction

The emergence of platform literature is shifting our thinking and prevailing organizational arrangements toward platform-centric entities, although this phenomenon has not yet received much attention in research. Furthermore, in accordance with, e.g., Schilling (2000) and Tiwana et al. (2010), we argue that the dynamics as well as the trajectories of these entities have remained understudied. In this chapter, we focus our analysis on understanding how platforms and their dynamics evolve through routine interactions.

L. Paavola (✉)
Department of Management Studies, Aalto University School of Business, Helsinki, Finland
e-mail: lauri.paavola@aalto.fi

© Springer Nature Singapore Pte Ltd. 2018

A. Smedlund et al. (eds.), *Collaborative Value Co-creation in the Platform Economy*, Translational Systems Sciences 11,
https://doi.org/10.1007/978-981-10-8956-5_7

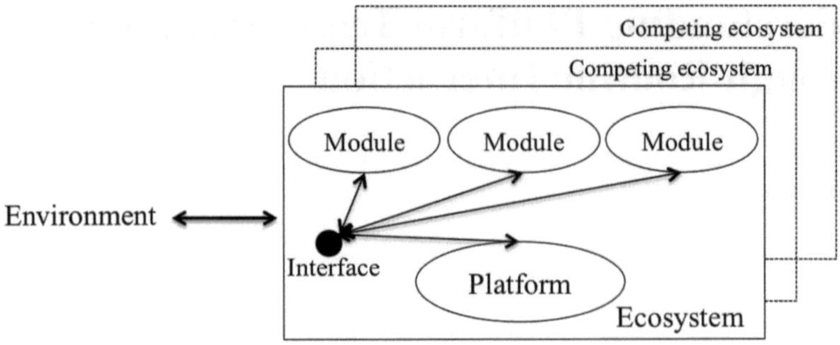

Fig. 1 Elements of platform-centric ecosystems (Tiwana et al. 2010)

Building on Baldwin and Woodard's (2009) synthesis on the commonalities across different platform-related conceptualizations, we define a platform as an extensible system to which modules that interoperate with the platform can be attached through interfaces (see Fig. 1). We define a module as an add-on subsystem that connects to the platform to add functionality to it (Baldwin and Clark 2000; Sanchez and Mahoney 1996). We refer to the collection of the platform and the connected modules as the platforms entity. The overall architecture – conceptual blueprint that describes how the entity is partitioned into a relatively stable platform and a complementary set of modules with predefined purposes (Katz and Shapiro 1994; Sanchez and Mahoney 1996; Ulrich 1995) – can create significant competitive advantages as well as barriers for rival platforms and ecosystems (Tiwana et al. 2010; Katz and Shapiro 1994).

We build our arguments on a notion that the operation of modules and platforms is determined by certain *routines* – scripts or codes of repetitive behavior – that they have in place for the execution of specific tasks and processes (Pentland and Feldman 2008; Ulrich 1995). Recently, scholars have focused on studying the properties and the evolution of individual routines mainly in organizational contexts (e.g., Feldman 2000; Howard-Grenville 2005; Cohen et al. (1996). On the other hand, an understanding of the properties of routines or their dynamics has never been applied to analyzing interfaces between modules and platforms. In this chapter, we make a case for the study of *transformational routines*, which we later define as scripts of processes by which modules or platforms either bring about changes or respond to them while interacting. We argue that an understanding of the emergence and transformation of modules and platforms can be obtained by reducing their analysis to the study of certain transformational routines, simultaneously providing insight into their broader multilevel change.

To better understand the progress of transformation, we also seek to broaden our discussion from transformational routines, the seeds of transformation, to how they advance multilevel transformation at different levels of platform entities as well as in their surrounding environments (cf. Gawer and Cusumano 2013). Due to the lack of literature on the dynamics of ecosystems, we consider it instructive to view the role of transformational routines through the discussions in institutional theory – another stream of literature that has had a significant impact on our current view on the origins and emergence of change in complex system. In their long-lasting debate on the origins of change, institutionalists have considered field-level change to involve both micro- and macro-levels.

We begin this chapter with a discussion of the nature and character of transformational routines. We follow this with an overview of the theory on the origins of change, first from the viewpoint of institutional theory and then through application to platforms, combining an understanding of the evolution of platforms and modules at the macro-level and the routine interaction at the micro-level. By combining our learning from these different streams of literature, we provide a synthesis and present a research approach for studying platform transformations at its different levels by means of transformational routines. Furthermore, we illustrate this learning through a case study from the field of UK grocery retailing where we analyzed the dynamics between a specific platform and a module (add-on subsystem) and finally discuss the methodological advantages that transformational routines as a unit of analysis offer for the study of dynamics of platform entities.

Defining Transformational Routines

Although the concept of routines has been in circulation since the early 1980s, research in the area has seen particular activity during the last decade. An organization's behavior can be considered to be determined by the routines it has in place for the execution of specific processes or tasks (Pentland and Feldman 2005). In the broad view, a given routine can exhibit a great deal of continuity over time, leading some theorists to emphasize their role in creating inertia and stability (Pentland and Feldman 2005). However, the research during the past decade has revealed that routines may change frequently and endogenously, and thus they can be viewed as a source of flexibility and change. The definitions of routines have developed and varied alongside the building of the theory. Perhaps the most suitable definition for our current understanding of a routine is the one by Cohen et al. (1996) who define it as an executable capability for repeated performance in some context that has been learned in response to surrounding pressures. Similarly, routines can be considered to determine the continuous operation and processes of both modules and platforms, which have also been considered as dynamic but adapting entities (Katz and Shapiro 1994). To emphasize the continuity and repetitiveness, we use the term *operational routines* (distinct from meta-routines and transformational routines, which we will discuss in the following).

A routine is considered to consist of two parts, termed the ostensive aspect and the performative aspect (Feldman 2000). The ostensive aspect represents the planned execution of the routine, whereas the performative aspect refers to the actual execution of the routine (Feldman 2000). Since the situations in which the ostensive processes are applied vary, the performative processes can differ according to the situational and environmental need for adaptation. The successes or failures of adapted processes are seen as a source of change to routines through learning (Rerup and Feldman 2011). Actors seek to repeat those adapted routines with

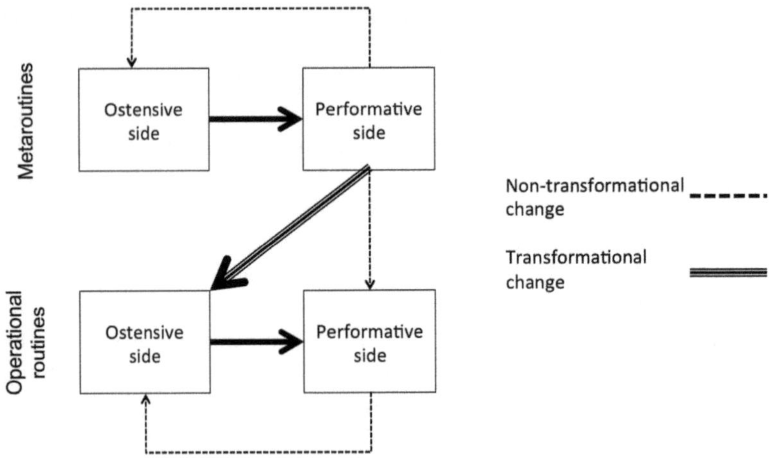

Fig. 2 Ostensive and performative aspects of routines

successful outcomes, further resulting in a change in their processes (Becker and Zirpoli 2008; Eisenhardt and Martin 2000).

However, such adaptation of routines, e.g., to a changing environment, is often insufficient in the long run. Change can itself be seen as a routine, which enables a module or platform to transform along with its operating environment by renewing or replacing its routines. We define a *metaroutine* as a routine that is designed to transform existing operational routines or to create new ones (Adler et al. 1999). Such a routine typically involves collecting, interpreting, and analyzing data concerning the way the operating environment changes. According to Weick (1995), it is insufficient to view an environment as a passive object that simply awaits to be discovered. Rather, any operator or observer needs to conduct action in order to understand what happens in its environment (Weick 1995). By means of adaptation processes, a module or platform may effectively select its environment from several alternatives and also influence it. Stinchcombe (1990) argues that the speed and contents of routines as well as switches made among them provide competitive advantage to certain operators. Consequently, systems in which modules and platforms adapt faster transform at a quicker pace.

In this chapter, we separate transformational routines from operational routines and define them as those metaroutines that lead to processes of profound change that orient a module or platform in a new direction and bring about changes in the ostensive aspect of its operational routines (Paavola and Cuthbertson 2016). Metaroutines can also be non-transformational, in cases where they only produce minor changes and function mostly in the role of a coordination mechanism that helps adjust the performative aspects of the operational routines (Paavola and Cuthbertson 2016). The possible influences of the various aspects of organizational routines on each other are summarized in Fig. 2.

Understanding Dynamics of Complex Systems

Micro-Macro Debate on the Origins of Change

Our current understanding of change in complex systems has been significantly shaped by institutionalists and their debate on the origins of change. Modern organizational theorists and institutionalists have started to acknowledge that institutions operate in an environment determined by other institutions. Early institutionalist works taking this perspective, similarly to the majority of theories on platforms, were mainly conducted from the macro-perspective. In particular, Meyer and Rowan (1977) analyzed macro-institutions like states and professions. These institutions are constructed by society and "become established as authoritative guidelines for social behavior" (Scott 2008). Inspired by these two pioneers, other so-called macro-institutionalists focused mainly on the attempts by organizations to achieve legitimacy by means of conforming to their macro-institutional environments (DiMaggio and Powell 1983). In particular, DiMaggio and Powell used the term *isomorphism* to describe similarity between the structures and processes of individual organizations. According to their account, organization-level and individual-level structures and processes emerge via top-down institutionalization stages.

A different view with more emphasis on the actions of individuals is provided by the so-called micro-institutionalists. Already Cyert and March (1963) as well as Nelson and Winter (1982) emphasized the importance of learning and the skills of individuals in organizational change. Moreover, according to Zucker (1977), Cohen (1991), and Feldman (2000), much individual behavior is governed by structures that originate and are regulated at the micro-level (Zucker 1977; Cohen 1991; Feldman 2000). Zucker (1977) argued that the generation of an institution starts from micro-processes at the intraorganizational level, offering a bottom-up account of institutionalization. She viewed the top-down institutionalization process merely as the reproduction of environments as institution – yet also this reproduction often occurs through the everyday activities of individuals (Zucker 1977). Micro-institutionalists agree with macro-institutionalists on institutionalization being a social construction as described by Berger and Luckmann (1966). However, micro-institutionalists assert that the social construction process leading to institutionalization begins at the micro-level, initially with two individuals interacting with each other. Also DiMaggio and Powell (1991) revised their neo-institutionalist approach, emphasizing micro-institutional processes and introducing practical action theory. Moreover, intraorganizational processes are not only important for institutional emergence, but "they can also be an important source of flexibility and change" (Feldman and Pentland 2003) as individual habits, performance programs, and genetic materials.

On the other hand, Suchman (1995) argued that qualifying institutionalization as a bottom-up or top-down process is not easy, since institutionalization can be driven by various actors at various levels of analysis. According to Suchman's model, the emergence of institutions depends on the demands set by unresolved problems

(Scott 2008). If there are ready-made solutions to problems, top-down institutionalization occurs through adoption, incorporation, or recombination of the existing solutions. This type of institutional isomorphism gives actors legitimacy (Dacin 1997; Deephouse 1996; Suchman 1995). Otherwise, actors construct new institutions via categorization, comparison of certain responses to problems, and development of alternative solutions. Leblebici et al. (1991) propose that any successful theoretical explanation and its empirical support must address the causes of a particular change in a field and explain how a novel practice eventually becomes standard for participants whose positions were previously sustained by the older practices.

Combining the Viewpoints to Understand Complex System Transformation

Research on change in platforms has traditionally focused on transformations in the overall ecosystem and its components (macro-level) or on individual processes and routines in a given component, i.e., a module or platform (micro-level). Yet, the majority of studies in this area take into account one of these levels only. It is demanding to combine the top-down and bottom-up viewpoints to platform transformations, as they involve very different units of analysis, producing challenges in the collection of data.

We argue that a combination of the analyses at the micro- and macro-levels would constitute a better understanding of transformation. Due to their properties, we consider transformational routines as a suitable unit of analysis in the study of module and platform transformations. Transformational routines operate as a medium of change among individual modules and platforms, binding their action together. We argue that the nature of change in platforms is determined by the operation of transformational routines that are in place.

To advance our knowledge of the complex cross-level processes involving routines (micro-level) and changes in modules and platforms (macro-level) as well as the ties between these, we seek to combine ideas from the literatures across the micro-macro spectrum, investigating the two levels and explaining the path dependencies between them. The concept of path dependence has previously been considered when analyzing individual actors, for example, by looking at how routines create competence traps, i.e., suboptimal procedures that are good enough in the short run, creating no need to introduce better procedures (Becker 2004; Levitt and March 1988). Transformational routines may help avoid competence traps. On the other hand, identifying the path dependence generated by routines at the macro-level highlights tensions and interference among, e.g., different parts of a dynamic setting. Hence the concept of routines also contains seeds to understanding conflicts and the compatibility of components of complex systems such as platforms.

Applying Transformational Routines as Units of Analysis in Understanding Platform Transformation

The previously presented literature suggests that analysis of transformational routines provides the building blocks for understanding transformations in complex settings such as platform entities. The literature indicates that transformational routines operate as links of interrelation between modules and platforms as well as the environment in which they operate. As the transformational routines form a critical link in the transformation process, the research approach that we suggest is founded upon their close analysis.

The transformation driven by transformational routines resembles what Van de Ven and Poole (1995) called the process of teleological change. In accordance with the philosophical doctrine of teleological change, transformational routines and their implementation are determined by the purpose and goal they are designed to accomplish (Van de Ven and Poole 1995). Furthermore, teleological theory typically views the development of an entity to be a process directed toward a goal (Van de Ven and Poole 1995). As in the social construction theory by Berger and Luckmann (1966), it is assumed that the entity is purposeful and adaptive. Through transformational routines, a module and platform evolves either by itself or in interaction with others toward an end state. A high level of creativity and localized processes are inherent in this type of process because different entities, consisting of individual units, have the freedom to pursue whatever goals they choose (Van de Ven and Poole 1995).

As the transformation process is guided by the end state, a suitable approach with a research approach focusing on transformational routines begins with targeting the data collection toward the transformations or evolution of certain functions of interest. After the choice of the area of analysis has been determined, it is possible to start an iterative process by identifying the environmental factors and other driving forces that have had an influence on this specific development. Through observing the actions that modules or platforms take in response to external stimuli, it is possible to identify the specific transformational routines to focus on. By iterating the process based on the phenomena of interest, a reduced view (Cohen and Harel 2007) of the driving forces and related routines can be developed. Furthermore, due to the repetitive nature of routines, a general understanding of the behavior and characteristics of modules and platforms can be created.

Application of the Suggested Approach in a Research Study on Use of Customer Data in Tesco

In the following, we present a research study in which we have applied the methodology involving the focus on transformational routines. In this case illustration, we first provide an overview of the analyzed development and then consider the existing theory and methodology on the analyzed platform. This is followed by a discussion where we consider the novel contributions and results produced by the use of transformational routines as a unit of analysis.

Customer Data Platform as a Driver of Transformation in Tesco

The evolution of the use of customer data in UK grocery retailing began in the early 1990s, when Tesco's marketing staff assessed retail loyalty schemes in the USA and concluded that a crucial benefit that could be obtained from such schemes was the accumulation of customer data. In Tesco's new card-based customer loyalty program, customers could sign up for their Clubcard, and by making purchases at Tesco, they would accumulate points that were converted into vouchers that could be used to purchase items at Tesco. This was intended to increase customer loyalty. Moreover, customers would give their names, addresses, as well as information on the size and ages of their family, and in this way, Tesco was able to identify the customer in each transaction performed at one of its stores.

During the first two months of the trial in 1993, Clubcard turned out to be very popular among Tesco's customers, but processing the massive amounts of transactional data with the limited computing power available at the time proved time-consuming and expensive. As a result, Tesco turned to a small consumer data analysis company named Dunnhumby, which showed that they could analyze around 10% of the available data and reach conclusions that were 90–95% accurate. This produced interesting findings on how far customers were willing to travel to shop, as well as which departments within stores failed to attract particular customers who shopped heavily in other areas. Moreover, previously Tesco had not known that a small proportion of customers produced a very large part of Tesco's profit.

The Clubcard was launched nationally in the beginning of 1995. In order to exploit the customer data, Tesco and Dunnhumby together created a platform for data sharing that allowed the findings to be operationalized throughout the organization. The modules that Dunnhumby created for the platform were efficient and included a large number of features, e.g., pricing, promotions, merchandising, and store development. All the data and its analyses resided on a server, so many different users could access the data at the same time. It had different levels of access and security for different kinds of users and parties. A number of different internal routines were implemented for the use of data: Dunnhumby sought to routinely update and develop/transform content of the data through their interface, whereas Tesco tried to find new and innovative ways for the actual implementation of the data.

Tesco's quarterly mailings of coupons were eventually tailored, in essence, to individual customers. Other uses of customer data took place at a broader level. Customer segmentation consisted of classifying customers into different groups, say, environmentally conscious buyers, bulk shoppers, etc. For example, it was found that one group of loyal customers regularly shopped at 12 of 16 Tesco store departments, implying that encouraging these customers to shop even sporadically in the other four departments would produce large profits. The segmentation also proved useful for Tesco in the price wars against competitors, especially Asda.

Another process related to customer segmentation was the so-called customer plan process. This is an annual process where the data was used to identify what is important to customers, what their opinions are of Tesco, etc. As a result, a small number of customer plan projects were conceived each year and put into effect. The next important step in the utilization of customer data was providing it to the supply chain partners. In the UK, and perhaps globally, this was pioneered by Tesco in 2002 when Tesco allowed its suppliers an access to their data platform simultaneously providing their own input on it.

Due to their highly successful collaboration, between 2001 and 2010, Tesco acquired Dunnhumby. As a result of this and the fact that data was now being routinely sold to interested parties such as the supply chain, the collection, analysis, and selling of customer data became a more mechanical operation. Furthermore, Dunnhumby was no more an independent partner that could pursue whatever goal they wished, and the company lost some of its drive to innovate new ways to analyze data. Whereas the routine that Dunnhumby was running had for years brought new innovative ways for looking at data and providing Tesco new tools for its use, the collaboration had now become more transactional. The routine that Dunnhumby was operating on the interface of Tesco's platform had shifted to primarily as non-transformation (see Fig. 2).

While many of Tesco's competitors, especially Sainsbury's, were trialing or launching their own loyalty cards simultaneously with Tesco, they did not place sufficient emphasis on the collection and use of customer data. More recently, between 2007 and 2013, most other grocery retailers in the UK have caught up with Tesco, partially because the analysis of data has been facilitated by advances in technology and computing power. On the other hand, with the advent of "big data," including customer data obtained from Internet sales, it is again impossible to analyze all available data, and so the various grocery retailers are competing on a rather even basis in driving innovations based on customer data.

Observations on Change Through Transformational Routines

In the study of Tesco, we focused on the way in which Tesco has been able to create new innovations with the help of customer data. Tesco's data-sharing platform (Fig. 3) has undergone a major transformation during the considered time period, with various routines acting as a vehicle for change. Such routines are jointly structured to reward buying behavior that is beneficial to the retail organizations, but even more importantly, they provide retail organizations with data on their customers that influences strategic and operational decision-making and may result in changes in organizational processes, involving pricing, promotions, merchandising, and store development. The overall idea was introduced in 1995 when Tesco launched its Clubcard and established a platform around the use of the data. As a

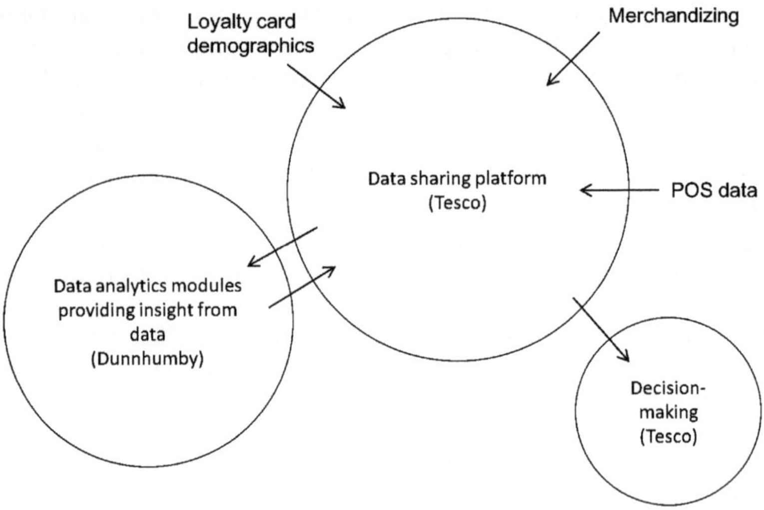

Fig. 3 Tesco's data-sharing platform

result of the benefits that Tesco gained from the launch, the competitive dynamics within UK grocery retailing have changed drastically, leading to efforts by other competing organizations to fight back.

Existing theories on the origins of change in platforms have previously focused on the conforming behavior on a certain specified level. By contrast, in our case, we showed that routines can in reality act in a dual role – in our case as both module- and platform-level units of analysis and drivers of field-level evolution, eventually leading to the launching and transformation of competing platforms.

We discovered that specific routines involving the collection and analysis of customer data have played a crucial role in the data-sharing platform of Tesco, as well as the evolution of the entire field of UK grocery retailing, and hence focused on these. In conducting our data collection, we organized interviews with certain key figures who had played a major role in developing the use of customer data in one or several of the organizations in question. We found out that while the ostensive aspect of a routine involving customer data would remain fairly stable, the incentive and drive with which the routine was executed would affect its performative aspect. In this study, we were able to understand the development of Tesco by analyzing routines that Dunnhumby was running to promote change. We were also able to understand the factors that first increased the influence of Dunnhumby (in the role of an add-on subsystem) on the platform governed by Tesco and similarly the factors that decreased its impact after being acquired.

Discussion: Methodological Advantages of Studying Transformational Routines in Module and Platform Transformations

As such, routines operate at the micro-level within modules and platforms. On the other hand, such components need to conduct action in order to select, assess, and adapt to their macro environments. This is achieved by means of transformational routines, which thus provide a means to better understanding both the localized and discontinuous pathways leading to variation among modules and platforms, as well as the general evolutionary processes (see Fig. 1). Due to their ability to provide insight into the nature of a complex system, the concept of transformational routines can also be applied to perform comparisons between different developments and the reasons behind them. In the following, we consider some of the advantages that these qualities of transformational routines provide in the study of transformations.

The concept of routines has been previously employed to understand organizational change by analyzing how organizational routines change (Nelson and Winter 1982). During the past decades, a large amount of research has been conducted based on the premise that routines are the basic components of behavior (Becker et al. 2005) and central in the accumulation, transfer, and application of capabilities (Cohen 1991; Cohen et al. 1996; Winter 2000). From this perspective, routines are a crucial part of any account of how a system accomplishes its tasks.

According to the properties of transformational routines observed in our study of UK grocery retailing, one of the main advantages of this concept derives from its ability to operate in a dual role as both an organization-level unit of analysis and a driver of platform evolution. At certain junctures in their development, external factors drive modules or platforms to trigger transformational routines that act as structuring mechanisms, and at others, transformational routines activate processes that drive module or platform evolution, transforming the cognitive, normative, and/or social structures of a complex entity. Thus there is an interplay between transformational routines, platform, and module evolution.

As noted in our literature review, existing theories of transformations tend to focus on continuous processes and static descriptions of historical events. Analysis of transformational routines provides an understanding of how a module or platform reacts, to what it reacts, and how the reactions evolve with respect to the environment, revealing the nature of a complex system as well as of the individual parties. These are difficult to observe with conventional methodologies and qualitative methods. Furthermore, since transformational routines bring about changes in routines themselves, they can be seen as a self-contained unit of analysis.

Existing theories of platform transformation remain largely on the macro-level. According to Becker et al. (2005), this type of analysis is incapable of capturing many interactions and their effects on actors and the environment. By contrast, transformational routines are micro-level units of analysis that provide tractable, temporally, and spatially limited settings to focus on while observing the environmental assessment and decision-making processes that bring about different levels of transformation. Hence the study of transformational routines is also close to the concept of scientific reductionism, which essentially entails reducing complex entities and interactions to the sum of their constituent parts, in order to make them easier to study. On the other hand, as Cohen and Harel (2007) note in relation with biological systems, understanding the essence of a system must include a consideration of its higher-scale emergent properties as well as of its fundamental component parts (such as transformational routines, in our setting) (Cohen and Harel 2007).

Due to the high level of focus involved, the concept of transformational routines enables the collection of data that can be fairly restricted in scope but still provide insight into module and platform, as well as the analysis of a complex system. As seen in our research study, possible methods include real-time collection of primary data, historical analysis of archival data, or combining both forms of data.

Conclusions

In conclusion, routine theory and institutional theory differ in their explanatory power of transformation in terms of their theoretical foundations, levels of analysis, and main arguments. We propose that combining ideas from these two theories by means of studying transformational routines enables a much richer understanding of transformations in complex systems, as one is able to link macro-level forces and actors with the situated actions and interpretations of micro-level actors. This is important when studying transformations that emerge as interplay between the two levels, as we saw in our case.

References

Adler, P. S., Goldoftas, B., & Levine, D. I. (1999). Flexibility versus efficiency? A case study of model changeovers in the Toyota production system. *Organization Science, 10*, 43–68.

Baldwin, C. Y., & Clark, K. B. (2000). *Design rules, Volume 1: The power of modularity*. Cambridge MA: MIT Press.

Baldwin, C., & Woodard, J. (2009). The architecture of platforms: A unified view. In A. Gawer (Ed.), *Platforms, markets and innovation* (pp. 19–44). Cheltenham: Edward Elgar.

Becker, M. (2004). Organizational routines: A review of the literature. *Industrial and Corporate Change, 13*, 643–678.

Becker, M., Lazaric, N., Nelson, R. R., & Winter, S. G. (2005). Applying organizational routines in understanding organizational change. *Industrial and Corporate Change, 14*, 775–791.

Becker, M. C., & Zirpoli, F. (2008). Applying organizational routines in analysing the behaviour of organizations. *Journal of Economic Behavior & Organization, 66*, 128–148.

Berger, P. L., & Luckmann, T. (1966). *The social construction of reality. A treatise in the sociology of knowledge*. New York: Doubleday.

Cohen, M. D. (1991). Individual learning and organizational routines. *Organization Science, 2*, 135–139.

Cohen, M. D., Burkhart, R., Dosi, G., Egidi, M., Marengo, L., Warglien, M., & Winter, S. (1996). Routines and other recurring action patterns of organizations: Contemporary research issues. *Industrial and Corporate Change, 5*, 653–698.

Cohen, I. R., & Harel, D. (2007). Explaining a complex living system: Dynamics, multi-scaling and emergence. *Journal of the Royal Society Interface, 4*, 175–182.

Cohen, M. D., Burkhart, R., Dosi, G., Egidi, M., Marengo, L., Warglien, M., & Winter, S. (1996). Routines and other recurring action patterns of organizations: Contemporary research issues. *Industrial and Corporate Change, 5*, 653–698.

Cyert, R. M., & March, J. (1963). *A behavioral theory of the firm*. Englewood Cliffs: Prentice Hall.

Dacin, T. M. (1997). Isomorphism in context: The power and prescription of institutional norms. *Academy of Management Journal, 40*, 46–81.

Deephouse, D. L. (1996). Does isomorphism legitimate? *Academy of Management Journal, 39*, 1024–1039.

DiMaggio, P. J., & Powell, W. W. (1983). The iron cage revisited: Institutional isomorphism and collective rationality in organizational fields. *American Sociological Review, 48*, 147–160.

DiMaggio, P. J., & Powell, W. W. (1991). *The new institutionalism in organizational analysis* (W. W. Powell & P. J. DiMaggio, Ed.). Chicago: University of Chicago Press.

Eisenhardt, K. M., & Martin, J. (2000). Dynamic capabilities: What are they? *Strategic Management Journal, 21*, 1105–1121.

Feldman, M. S. (2000). Organizational routines as a source of continuous change. *Organization Science, 11*, 611–629.

Feldman, M. S., & Pentland, B. T. (2003). Reconceptualizing organizational routines as a source of flexibility and change. *Administrative Science Quarterly, 48*, 94–118.

Gawer, A., & Cusumano, M. (2013, September 4). Industry platforms and ecosystem innovation. *Journal of Product Innovation Management, 31*, 417–433.

Howard-Grenville, J. A. (2005). The persistence of flexible organizational routines: The role of agency and organizational context. *Organization Science, 16*, 618–636.

Katz, M., & Shapiro, C. (1994). Systems competition and network effects. *Journal of Economic Perspectives, 8*(2), 93–115.

Leblebici, H., Salancik, G. R., Copay, A., & King, T. (1991). Institutional change and the transformation of interorganizational fields: An organizational history of the U.S. radio broadcasting industry. *Administrative Science Quarterly, 36*, 333–363.

Levitt, B., & March, J. G. (1988). Organizational learning. *Annual Review of Sociology, 14*, 319–340.

Meyer, J. W., & Rowan, B. (1977). Institutionalized organizations: Formal structure as myth and ceremony. *American Journal of Sociology, 83*, 340–363.

Nelson, R. R., & Winter, S. G. (1982). *An evolutionary theory of economic change*. Cambridge, MA/London: Belknap Press of Harvard University Press.

Paavola, L., & Cuthbertson, R. (2016). *Redefining metaroutines as drivers of transformation: Use of customer data in food supply management*. Academy of Management Proceedings 2016.

Pentland, B. T., & Feldman, M. S. (2005). Organizational routines as a unit of analysis. *Industrial and Corporate Change, 14*, 793–815.

Pentland, B. T., & Feldman, M. S. (2008). Designing routines: On the folly of designing artifacts, while hoping for patterns of action. *Information and Organization, 18*, 235–250.

Rerup, C., & Feldman, M. S. (2011). Routines as a source of change in organizational schemata: The role of trial-and-error learning. *Academy of Management Journal, 54*(3), 577–610.

Sanchez, R., & Mahoney, J. (1996). Modularity, flexibility, and knowledge management in product organization and design. *Strategic Management Journal, 17*(1), 63–76.

Schilling, M. (2000). Toward a general modular systems theory and its application to interfirm product modularity. *Academy of Management Review, 25*(2), 312–334.
Scott, W. R. (2008). *Institutions and organizations: Ideas and interests*. Thousands Oaks: Sage.
Stinchcombe, A. L. (1990). *Information and organizations*. Berkeley: University of California Press.
Suchman, M. C. (1995). Managing legitimacy: Strategic and institutional approaches. *Academy of Management Review, 20*, 571–610.
Tiwana, A., Konsynski, B., & Ashley, A. A. (2010). Coevolution of platform architecture, governance, and environmental dynamics. *Information Systems Research, 21*(4), 675–687.
Ulrich, K. (1995). The role of product architecture in the manufacturing firm. *Research Policy, 24*, 419–440.
Van de Ven, A. H., & Poole, M. S. (1995). Explaining development and change in organizations. *Academy of Management Review, 20*, 510–540.
Weick, K. E. (1995). *Sensemaking in organizations*. Thousand Oaks: Sage Publications.
Winter, S. G. (2000). The satisficing principle in capability learning. *Strategic Management Journal, 21*, 981–996.
Zucker, L. G. (1977). The role of institutionalization in cultural persistence. *American Sociological Review, 42*, 726–744.

Creating the Foundation for a Functioning Internal Platform

Mikko Hänninen, Olli Rusanen, and Lauri Paavola

Abstract This chapter provides an empirical example of how platform theory can be used to study organizational networks. Platform theories have recently received increased attention from academics, as digitalization has fueled the implementation of digital platforms and platform-based business models across a wide range of industrial settings. However, despite the increased interest toward platforms, most of the research is still confined to the economics and engineering schools of thought. In this chapter, we address this theoretical and empirical gap by providing an empirical example of a retail cooperative that can be characterized and analyzed as an internal platform. We study how this type of organization is managed, focusing on the change process through which the organization shifted from a crisis organization in the early 1980s to an efficient platform-like organization by the mid-1990s. By comparing the organization before and after changes in its organizational structures and practices, we identify the steps taken to make the most use of a platform-like organizational structure. We identify three building blocks created by the central organization that provided the foundation for the cooperative's future success: the divestment of value-destroying and non-valuable resources, preventing exploitative use of resources, and enabling participants to identify and create new activities. Through our case study, we provide future research avenues to the internal platform stream of platform literature and invite empirical research that applies platform literature to different contexts, for example, to looking at platforms also as an organizing and organizational form.

Keywords Platform · Internal platform · Decentralized organization · Retail · Strategic management

---

M. Hänninen (✉) · O. Rusanen
Department of Marketing, Aalto University School of Business, Helsinki, Finland
e-mail: mikko.o.hanninen@aalto.fi

L. Paavola
Department of Management Studies, Aalto University School of Business, Helsinki, Finland

© Springer Nature Singapore Pte Ltd. 2018

A. Smedlund et al. (eds.), *Collaborative Value Co-creation in the Platform Economy*, Translational Systems Sciences 11,
https://doi.org/10.1007/978-981-10-8956-5_8

Introduction

During the past few years, platform theories have received notable attention and expansion from academics as digitalization has spawned the creation of multisided platform-based business models across a wide range of industries (e.g., Gawer 2014; van Alstyne et al. 2016; McIntyre and Srinivasan 2017). At the heart of platform thinking is a collaborative network where a platform provides an interface that is open for external complementors to create applications to (e.g., Adner and Kapoor 2010; Autio and Thomas 2014). Firms such as Apple, Facebook, and Google are often used as examples of firms that orchestrate a platform ecosystem that allows for value creation and innovation within a network of firms or complementors (e.g., van Alstyne et al. 2016). However, due to strong influence of the economics and engineering schools of thought, platform research has so far studied extensively price setting and network externalities as well as technological infrastructures (e.g., Gawer 2014; Gawer and Cusumano 2014; Tiwana 2014), leaving the intraorganizational, internal platform context with less theoretical and empirical advances. Despite Gawer's (2014) definition of an internal platform, empirical studies in the internal platform context are still largely missing, especially in terms of using the platform theory and the internal platform stream of platform literature, to analyze the functioning of decentralized and network organizations.

Regardless of the type of platform, value creation in platform ecosystems takes place in the interaction of platform participants. A platform ecosystem is a network of participants consisting of the platform owner, complementors, and users (e.g., McIntyre and Srinivasan 2017) with a platform's role defined as facilitating the interaction between these different user groups. Thomas et al. (2014) describe platforms through architectural leverage where platforms are designed to achieve high impact with little input from the platform owner in order to generate high profits for platform owners. The question of management of an internal platform is particularly interesting and not fully addressed in the existing platform literature outside the product development and supply chain contexts (e.g., Gawer and Cusumano 2014; Thomas et al. 2014). In this chapter, we show that organizational networks can be analyzed as internal platforms, as these types of organizations share several characteristics to platforms. For example, the headquarters orchestrates value in organizations, by providing the modules on top of which the organizations' other stakeholders can create value added to.

This chapter contributes to strategic management and platform literature by showing that organizational networks function according to the logic of internal platforms. By studying an organization as an internal platform, we take the first step in providing empirical examples of internal platforms outside the product development context with the internal platform structure relevant for other types of organizations as well, such as organizational networks and other decentralized organizational forms. We take a non-digital perspective to platforms such as other internal platform papers have done in the past (e.g., Simpson 2004) and use the Gawer (2014) framework to analyze our single case study organization, retail cooperative S Group, as an internal platform. In our case study, we focus specifically on the management of such a network organization that functions as a platform and answer the question of how the management can make the most use of such a platform structure. The coordination and management of network organizations, like cooperatives, have traditionally posed several questions for academics (Provan and Kenis 2008), and this paper is the first to use the internal platform framework to study the functioning of an organizational network. Based on the case study, we identify three building blocks that in our single case example allowed the cooperative to maximize the benefit of a platform-like organizational structure and find efficiency throughout the organization.

We proceed by providing a short overview of recent advances in platform literature, followed by a review focusing on the internal platform stream of literature. We then justify the methodological choices applied in our single case study, followed by the case study of a retail cooperative from the retail sector which we define to function as an internal platform. Finally, we discuss the key findings from our case study and provide future research avenues to help expand the internal platform stream to topics such as organizational and strategic management.

Theoretical Background

In this section, we justify why platforms and specifically internal platforms are an interesting research area, focusing on identifying the key characteristics of platforms that can be applied to organizational networks. We shortly review the recent developments in platform literature and then focus specifically on the internal platform stream which we seek to expand through our single case study where we study a retail cooperative as an internal platform.

The Engineering and Economics Perspective to Platforms

Platform theory originates in the engineering and economics theoretical schools of thought. This stream of literature defines platforms as dynamic and purposive internal or external networks (e.g., Adner and Kapoor 2010; Autio and Thomas 2014). In these types of networks, the success of a platform depends on the ability for platform participants to create value through their interactions with other participants (e.g., Tiwana 2014). In the more traditional definition of platforms, for example, in the context of new product development, the role of the external or internal platform participants is to develop complementary products, services, or technologies on top of the platforms' standardized modules (e.g., Jacobides et al. 2006; Gawer and Cusumano 2014). Platform owners such as the technology providers simply orchestrate these interactions between the internal or external complementors and the platforms' own digital or physical infrastructure. In the context of internal platforms,

platforms are built around modules on top of which supply chain members can create their own offering, while in digital platforms, modules are replaced by technology interfaces.

Most research on platforms has focused on the so-called industry platforms. In an industry platform, the platform owner has given up control of the platform's components to independent complementors in order to allow the creation of complementary components and assets (Thomas et al. 2014). Examples of industry platforms include the Apple iTunes Store, where the iTunes platform serves as the interface to which external complementors (app developers) can built their own content (applications) according to the rules and guidelines set by the platform owner which in the iTunes Store example is Apple (Tiwana 2014). Overall, the recent academic literature around these types of industry platforms has moved from studying topics such as pricing (Armstrong 2006) to covering broader areas related to platform management such as governance and competitive strategy (Tiwana 2014).

The recent interest on platforms has emerged through the advent of digitalization. In the context of digital services, academics often talk about multisided markets (e.g., Hagiu and Wright 2015) with companies like Uber and Airbnb as the face of this digital platform revolution as it is popularly referred to as (e.g., van Alstyne et al. 2016). A multisided market means that a platform enables the interaction between two or more sides of participant groups. Thus, in an external two-sided platform, the main purpose of the platform owner is to maximize the network effects (Eisenmann 2006; Haucap and Heimeshoff 2014) and minimize entry barriers, for example, through the platform design and architecture (Eisenmann et al. 2011). The platform's value is based on the number of users, and the core assets of the platform owner are the platform's participants such as the customers and service providers. The platform owners thus face a chicken-and-egg problem in these types of two-sided markets; it must attract a large base of service producers and customers simultaneously, and the value for this user base is based on their mutual interaction (e.g., Gawer and Cusumano 2002; Rochet and Tirole 2003). This creates a dilemma as the switching costs for platform participants are low and there may only be loose contractual relationships toward other participants of the platform.

The Internal Perspective to Platforms

Compared to industry platforms, internal platforms are a literature stream of their own (e.g., Thomas et al. 2014). Thus far, this internal perspective to platforms has received only limited empirical interest from scholars across disciplines. For example, the internal platform framework by Gawer (2014) has received less support than her other works, for example, on industry platforms. So far, literature on internal platforms has primarily focused on the context of new product development and engineering as we outline in this section.

In the product development context, a product platform is defined as the set of common components, elements, or assets that can be shared across the organization (e.g., McGrath 1995; Meyer and Lehnerd 1997; Robertson and Ulrich 1998). Simpson (2004), for example, introduces two types of product platforms, module- and scale-based product families. Popular examples of successful product platforms include Sony Walkman and Black & Decker (e.g., Gawer 2014) in the consumer electronics and machine tools industries, respectively. The Sony Walkman product portfolio, for example, was built around key modules that allowed Sony to utilize the modular design and flexible manufacturing processes to introduce a large product portfolio, spanning over 250+ models in the 1980s (e.g., Sanderson and Uzumeri 1996). On the other hand, Black & Decker is often used as an example of a successful scale-based product platform, developing in the 1970s a family of universal motors for power tools that varied only with regard to stack length and in the amount of copper wrapped around the motor, allowing it to scale the standardized motor for the different Black & Decker product divisions.

In this chapter, instead of talking about internal platforms as product platforms in the product development and engineering context, we study how internal platform literature can be incorporated to study organizations, for example, utilizing the frameworks by Gawer (2014) and Thomas et al. (2014). Furthermore, we focus on understanding the functioning of such an internal platform that consists of one focal organization and its ecosystem of network members. Although research has looked at the internal structure of platforms (e.g., Cusumano and Gawer 2002), only a few definitions for internal platforms exist. The key commonality between the different types of platforms is that all platform participants contribute to the total value of the platform and co-create value with other participants regardless of whether the platform operates only inside one focal organization such as a product development platform or is open for external complementors such as an industry platform.

Gawer (2014) categorizes the key differences between industry platforms and internal platforms as structure and control. Internal platforms are inside the boundaries of a firm rather than organized as an ecosystem and are closed rather than open to external complementors (e.g., Gawer 2014). Organizations can adopt such an internal platform structure to respond to a rapidly changing external environment where the organization needs to efficiently generate new combinations of resources, routines, and structures (Ciborra 1996). Although literature defines industry platforms and internal platforms as two separate types of platforms, Gawer (2014) argues that it is possible for an internal platform to evolve to an industry platform as is evident through the case of IBM where an ecosystem of PC manufacturers emerged from IBM's supply chain which resulted in the eventual demise of IBM's, at the time, market-leading PC division. In the case of IBM, members of IBM's supply chain like Intel and Microsoft began to embrace the platform structure shifting from only operating in a closed one-firm supply chain of IBM to an industry platform which by the late 1980s already included an abundance of PC manufacturers.

Platforms are an interesting way to organize business from the intra-firm internal platform perspective as well, as platforms can be both fixed or able to change over time. These types of organizational networks require a strong role from management.

Cusumano and Gawer (2002), for example, state that platform owners need to manage constantly both external and internal conflicts taking place in the platform ecosystem. In the case of an internal platform, by management we refer to the headquarters. A platform is generally linked to an organization's headquarters that is in charge of orchestrating the value generated across the organization (Baldwin and Woodard 2008). The modules in such platforms can consist of, for example, standardized processes created by the headquarters such as standardized accounting practices and brand concepts. Overall, there is thus a direct link to strategic management literature, where the value generated by the headquarters is generally named as one of the four core problems in strategic management (Rumelt et al. 1994). The use of an internal platform framework can thus help create new theoretical advances also in strategic management literature.

Methodology

In this paper, we move away from product platforms that have traditionally been at the heart of internal platform research (e.g., Sanderson and Uzumeri 1995; Simpson 2004) to study S Group, a Finnish cooperative retailer, as an internal platform. As studies incorporating platform theory to study organizations are nonexistent (for an exception, see Ciborra 1996), we follow the inductive methodology of Burgelman (2011) and construct a conceptualization of organizations as internal platforms that applies to decentralized and network organizations.

Research Design

This chapter is built upon a single case analysis (e.g., Yin 1994) with the qualitative analysis using coding to analyze the plentitude of data sources (e.g., Gioia et al. 2013). Gioia analysis is used as the coding technique for creating theory that emerges from the data and showing the results visually. In such an analysis, the researchers aggregate data to first-order concepts, which are then grouped to second-order themes. Finally, the second-order themes are combined as theoretically relevant, aggregate dimensions (Gioia et al. 2013). As Burgelman (2011) argues, theory generation requires avoidance of theoretical preconceptions and constant comparison, coding, and analysis between theory and data. A Gioia analysis is helpful for achieving these ends since it aggregates raw data into theoretically relevant findings.

A Gioia analysis generally includes data tables (Corley and Gioia 2004) which show how first-order categories were created from the data. We have selected to show these data tables in narrative form, as a single case narrative increases the accuracy of the results while not being simple nor general (e.g., Langley 1999). This enables us to bypass the pitfall of Gioia analysis where the process that generates the aggregate dimensions is lost. The strength of Gioia analysis is that it presents findings in a simple table, so these two approaches are in fact complementary.

In this chapter, we aim to understand the management of a retail cooperative, S Group, between 1983 and 1996 and the steps taken by the organization's management to maximize the efficiency of the organization that we characterize as an internal platform. We focus on this period when significant changes happened in the organization that helped S Group maximize the benefit of its cooperative business model and allow the headquarters, central organization SOK, to take a more active role in the management of the organization as a primarily support function for the independent cooperatives across Finland. In terms of generalizability (e.g., Eriksson and Kovalainen 2008), the purpose of this chapter is to provide a better understanding for practitioners and academics of the possibilities provided by incorporating platform theory to organizational and strategic management studies, raising important theoretical questions in the process that should be addressed by further studies, especially about how organizations can make the most use of such a complex platform-like organizational structure.

Data Collection and Case Selection

We collected data through a longitudinal case study about the management of the S Group between 1983 and 1996. Collecting data from a single case is suitable to the topic as it allows us to gain an in-depth understanding of the phenomena and develop new constructs (e.g., Suddaby 2010). S Group was chosen as the case as it is a decentralized organization, a unique organizing form, and thus gives us the opportunity to study the applicability of platform theory to studying organizational networks through an extreme case example (e.g., Flyvjberg 2006; Siggelkow 2007). We analyzed a large amount of data and interviewed key figures in the organization in order to gain an in-depth understanding for the status quo at S Group in 1983 and the transformations that took place between 1983 and 1996. Over the course of the entire research process, we actively studied publicly available material on S Group in order to come up with a list of potential interviewees and an understanding of the context the cooperative was in at the time. We were also granted access to the cooperatives central organization (SOK) archives with our focus on all the past top management meeting minutes, quantitative graphs, and numerical figures describing strategic plans or the state of the company as well as a plentitude of documents, memorandums, transcripts of meetings, copies of old contracts, photographs, and other notes.

As a primary data source, we used semi-structured interviews. Interviews were organized with top management, middle management, and cooperative management. The informants included all the six SOK CEOs since 1983 and other individuals who had taken part in strategic decision-making between 1983 and 1996. Although we relied on the interviews to give us an inside view into the organization, we also used archival data during the research process in order to achieve maximum

Table 1 Description of primary and secondary data sources

Data	Type of data	Amount and description of data sources	Public/ confidential
Company archives	Archival data	Board minutes, contracts, personal memos 1980–1996	Confidential
In-depth interviews	Informants	6 CEOs of SOK	Confidential
		Top and middle management of SOK and cooperative management	
CEO archives	Archival data	Private notes, memos, photographs, presentations, newspaper clips 1983–2002	Confidential
Annual reports	Public data	Annual reports 1970–1996	Public
Books	Public data	10+ books or biographies on case company	Public

accuracy and legitimacy. The selection of interviewees was iterative, as informants also provided further information about key decision-makers throughout the interview process. The interviews were semi-structured, with a duration of 60–250 min. All interviews were taped, transcribed verbatim, and thoroughly analyzed by the researchers. The interviews were conducted and transcribed in Finnish, the native language of the informants. The data used in the study is described in more detail in Table 1.

This broad data set allowed us to compare the top and middle management views. We triangulated all main events from multiple sources, for example, all interviews were triangulated with archival data and vice versa. The interviews were conducted until saturation was reached.

Based on the empirical material, we created an event database to cover the period between 1983 and 1996 using both qualitative and quantitative data. These databases form the basis for the data analysis as they allow us to pinpoint key managerial decisions taken as well as their direct and indirect effects on the organization. These events correspond to what Gioia et al. (2011) describe as first-order categories. We also have temporally orchestrated data prior to the period of intensive analysis (pre-1983) in order to understand the research context also during the period before the change process in the organization was initiated. We also collected data covering after the period of intensive analysis (post 1996) in order to understand the position that the company was in after this process. This corresponds to the notion that in process studies, the end is the starting point in a continuous process (Langley et al. 2013; Tsoukas and Chia 2002).

Overview of the Single Case: S Group

S Group is a Finnish customer cooperative, which has since mid-2000s been the market leader in the Finnish grocery retail market with an over 40% market share. The organization consists of the central organization (SOK), independent cooperatives, and subsidiary operations which together form the S Group. Our analysis focuses on the transformation of S Group in the 1980s and the actions taken by the central organization SOK to create change toward more efficiency. While in 1983 S Group was a diversified company with operations in several industries from retail to agriculture, since the mid-1980s the grocery retail business has been the flagship industry, and other businesses have been divested, or their role in the business portfolio has been greatly diminished.

Coming into the 1980s, S Group was in deep crisis, and our data shows that the organization was in the brink of bankruptcy if immediate reforms were not initiated. We identified several causes for the crisis, but the most important one was a structure that enabled suboptimization as each cooperative tried to maximize their own position at the expense of group-level performance. Our data shows that the position of S Group in the 1980s had become severe due to a lack of systematic group-level internal accounting practices and diversification of S Group's business portfolio. Thus, as a result of these issues, little information about the financial situation or business performance flowed from the central organization to the cooperatives and vice versa. In 1983 S Group eventually turned to an external CEO, the first in their history, with a mandate to initiate reforms to turn around the company.

Today, largely due to the reforms done in the 1980s, the central organization SOK is only a support function for the regional cooperatives in Finland. Customer orientation is built through interaction and shared activities such as national retail chain across the S Group. In our empirical study, we focus specifically on identifying the steps taken by S Group to maximize the efficiency of its platform-like organizational structure. Through the restructuring in the 1980s, SOK and its regional cooperatives developed several customer-oriented initiatives that formed the competencies where S Group rose to become the market leader of the Finnish grocery retail market. In 1996, the major part of these key initiatives had been put in place, which is also the end of our study.

Case Study: S Group

S Group is a retail cooperative consisting of a central organization, SOK, and independent cooperatives that together form the S Group. Like the rules of organizational networks (e.g., Provan and Kenis 2008), S Group is not a legal entity as each member of the cooperative is autonomous and decentralized in their decision-making. In this paper, we characterize S Group as an internal platform, where the internal platform participants, the independent cooperatives, can share and use a common activity base consisting of, for example, chain and brand concepts, thus resembling an internal platform-like structure. Based on Gawer (2014), the S Group platform consists of an ecosystem of subunits, the regional cooperatives. The coordination of the platform happens through the central organization SOK, which has managerial authority according to the rules of the cooperative.

The internal platform-like structure was created at S Group in the early 1900s to ease the coordination between regionally diversified cooperatives that, without the platform, the central organization as a mediator would not be able to achieve economies of scale and thus be profitable. The platform structure was thus needed to create collaboration between Finnish regional cooperatives which in the late 1900s each ran only a small portfolio of stores in their region. This platform-type organization structure had existed throughout the history of the cooperative, but over the years it had become inefficient. Rather than a change of organizational structure, SOK management adopted several new practices between 1983 and 1996 that drastically increased the efficiency of the organization. Figure 1 shows the S Group

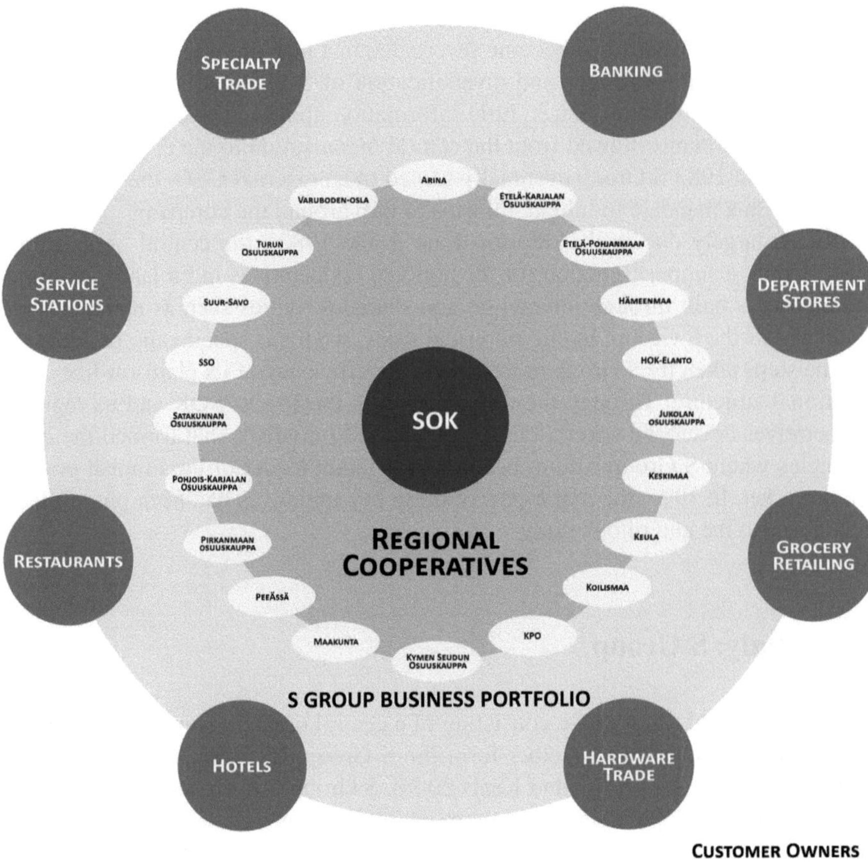

Fig. 1 S Group ecosystem

ecosystem as of today, after the transformation and structural change in the 1980s and 1990s, with 20 regional cooperatives and a large business portfolio run by the cooperatives.

Our analysis shows that S Group went through four crucial phases that allowed the cooperative to function more efficiently. We identified four phases in our event data, (1) status quo and (2) resource divestment, (3) platform orchestration, and (4) resource discovery and expansion, that allowed the organization to transform as shown through the Gioia analysis results in Fig. 2. It is noteworthy that typically Gioia analysis does not contain an internal structure between aggregate dimensions; as the time-coding of major strategic events shows, the aggregate dimensions of our analysis have a flow from 1 to 4. The following narrative describes how S Group shifted from a crisis organization in 1983 to finding efficiency through a few key strategic initiatives, opening up the key results from the Gioia analysis.

Status Quo at S Group (1983)

Coming into the early 1980s, S Group was a stagnant and complex organization that had poor resource allocation and information asymmetries due to a lack of standardized group-level practices. Customers perceived S Group's grocery stores as expensive and old-fashioned compared to competing retailers such as then market leader K Group. Inefficiencies were caused primarily by diversified decision-making in the S Group, as each individual cooperative oversaw running and planning the business of its own cooperative, in a certain geographical area. Due to the decision-making inefficiencies, each unit of S Group had a large workforce working for it, for example, each had their own human resources and marketing departments.

Due to numerous exploitative financial practices, the financial position at S Group was also weak in the early 1980s. S Group supported weak cooperatives financially, and there was a so-called internal monopoly in place where each subunit at S Group aimed to maximize their own unit's profits rather than the total value of S Group. Each unit in the cooperative, including the central organization SOK, was allowed to make a profit, meaning that there was a lot of suboptimization in place. As a result of an unrealized strategic plan that had proposed a restructuring of S Group already in 1969, there was also a lack of strategic direction that continued throughout the 1970s. All of this meant that several value-destroying and nonvaluable resources existed in the S Group at the time, and the entire organization was at the brink of bankruptcy.

In order to turn around the company, SOK brought in their first external CEO in 1983. The new CEO started a period of divestment of noncore businesses once the SOK management realized the true financial state of the organization. Through a strategic plan, S83, SOK established a regional cooperative structure at S Group, in which the previous network of 170+ local cooperatives was consolidated into a network of 36 regional cooperatives, by, in practice, forcing the mergers of smaller

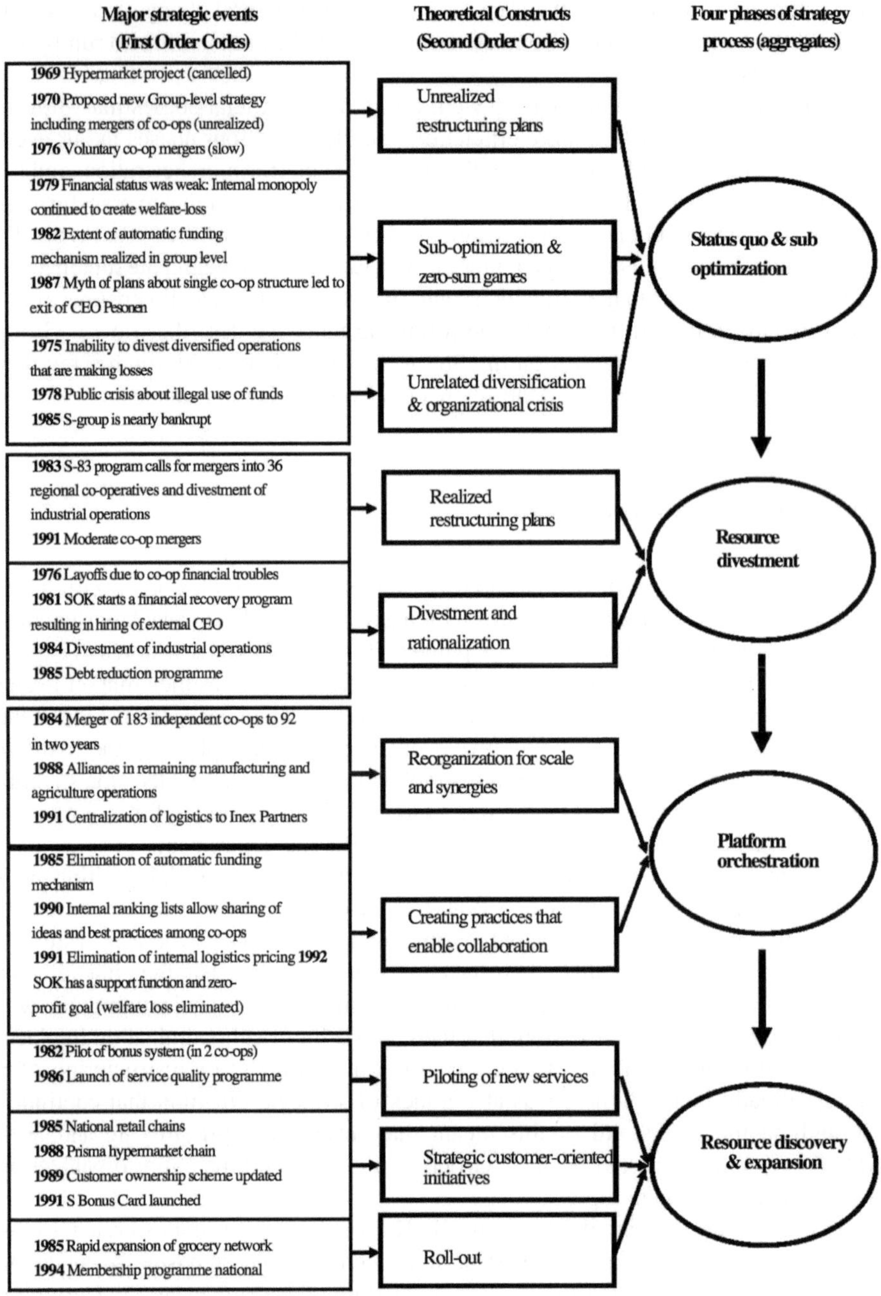

Fig. 2 Results of the Gioia analysis

cooperatives with their regional neighbors. Within S Group, cooperative mergers are voluntary and require acceptance of each cooperative, as they are independent businesses also with their own independent decision-making bodies. A crucial way to pressure the cooperatives to merge was to cut the central organization's financial support from the cooperatives. This shifted the crisis toward cooperatives who received advice and support only on the condition of mergers and divestment of unprofitable businesses. The restructuring also allowed SOK to open the balance sheets of struggling cooperatives in order to realize fully the catastrophic financial state of S Group. Significant losses were made until the late 1980s as the financial condition of several cooperatives was extremely poor due to, for example, a lack of amortizations on property and machinery.

Through the restructuring launched by the new CEO in 1983, S Group began to focus on the grocery retail business where the market share of S Group had been steadily falling over the past decade. The decision to focus on grocery retail resulted in the divestment of diversified businesses such as the agriculture and industry businesses throughout the 1980s and early 1990s. This led to layoffs and the gradual replacement of cooperative CEOs and SOK top management with recent business school graduates, who had more expertise in strategy work and were not guided by any existing ways of thinking. The majority of S Group top management in 1996 had come into the organization during the worst crisis years in the early and mid-1980s.

Transformation of S Group to a Functioning Internal Platform

The transformation started from the internal reorganization at S Group that was initiated from 1983 onward. The previous structure with hundreds of cooperatives had resulted in the lack of group-level thinking as instead of all stakeholders trying to maximize the value of the total S Group, each S Group unit and subsidiary tried to maximize their own unit's profits or their personal benefit. This gradually changed through the late 1980s with the introduction of regional cooperative structure and the elimination of past practices that had resulted in suboptimization.

The divestment of noncore businesses and the restructuring of purchasing functions were important for increasing internal efficiency. By 1992, S Group had reorganized its purchasing functions by centralizing its purchasing to its newly established subsidiary Inex Partners and divested its agricultural and industrial businesses to joint ventures such as Meira Oy. This divestment of diversified businesses shifted more roles for the planning of concepts to SOK, with the cooperatives now responsible for only operating their businesses according to the shared common vision and strategic goals. For example, until the mid-1980s S Group did not have a national retail chain network or standardized store concepts across the country.

The establishment of Inex Partners Oy and the changes in organization practices throughout the 1980s were important because they defined the cooperatives as profit centers and SOK now as simply a support unit. In the case of S Group, the "standardized modules" found in internal platform literature consisted of, for example, standardized accounting practices, standardized store concepts, and managerial guidelines. By 1991, internal logistics pricing was eliminated, and the cooperatives could source products directly from Inex Partners without shuffling products through internal middlemen in the S Group. All profits made by Inex Partners were channeled back to the cooperatives at the end of each year, according to the ratio of purchases made by each cooperative. This ensured that S Group remained competitive and that each cooperative was treated equally without the internal monopoly structure that had existed before.

As S Group began to maximize the benefit of its decentralized structure, it could create new activities especially guided by the middle management and the regional cooperatives. While previously some cooperatives had run a lot of independent pilots since the 1970s, such as hypermarkets and the purchase refund system (which later became the S Bonus Card), now because of the steps taken by the management during the 1980s, information was more actively shared across S Group, and several concepts were quickly rolled out nationally after local pilots. For example, in 1988 S Group launched the Prisma hypermarket chain nationally and in 1991 the S Bonus Card, both of which had originated from the cooperatives and later became ventures managed by SOK. Once the middle management participation in decision-making was secured, the rollout of new concepts such as national grocery store concepts was relatively fast.

While already in the 1970s S Group had skilled employees both in SOK and the cooperatives, there had previously been little intrinsic motivation for an individual employee or business unit to develop or turn around the business. This changed in the mid-1980s through the start of a culture of piloting new concepts and by giving middle management a stronger role in strategy planning and implementation. For example, in the mid-1980s initiatives such as the creation of national retail chains was led and implemented by middle management. The national retail chains were important for the restructuring of the grocery store network and in order to make sure that regardless of the location of the store in Finland, the store concept would be almost identical for the end-customer. The new middle managers that had come into the organization in the early 1980s were also responsible for introducing new practices such as the internal ranking lists of cooperatives, which ranked the regional cooperatives from best to worst and helped create healthy internal competition. This meant that there was also internal pressure from the other cooperatives to develop their business in order to make sure that their ranking was higher and that all cooperatives did everything they could to maximize the total value of the organization. While in the 1970s, the cooperatives were playing a zero-sum game due to financial support from the central organization, through the adoption of internal ranking lists, the cooperatives had means to identify best-performing operations and transfer and expand the processes that were valuable. In short, the internal ranking lists enabled internal imitation of processes across the organization.

The most successful initiative that came with the reforms was the renewed focus on the customer owners in the late 1980s and the subsequent launch of the membership card, S Bonus Card, in 1991. Our data shows that over the years, S Group had neglected to develop customer schemes and operations that would differentiate it from its competitors and deliver value for customers – which resulted in a continuously declining customer owner base. The focus on customers was restored through a strategic initiative launched in 1986, in which S Group redefined its mission to be about delivering superior advantages and benefits to customer owners as a customer cooperative such as S Group should exist only for its customers. The main action taken to implement this strategy was the launch of the bonus system and membership card in 1991 which had throughout the 1980s been independently piloted in a few regional cooperatives. Through the S Bonus Card, customers received purchase refunds based on their purchases in a system that increased progressively, i.e., rewarding the best customers the most for their purchases, up to 5% per month. After several pilots the membership card and bonus system were rolled out nationally in 1996 across all S Group businesses also outside grocery retail.

Discussion

Through the single case study, we provide an empirical example of an organization that can be defined as an internal platform and how such an organization can be managed to make the most use of such a platform structure. As there is no fixed definition for platforms (e.g., Gawer 2014), platform logic can be applied to organizations from several different contexts and industries. This chapter is one step toward extending the internal platform stream of literature to also cover platforms outside of the product development and engineering contexts and using the frameworks to study the functioning of organizational networks. We also generate new understanding and further research opportunities on internal platforms especially from the managerial point of view. Through a single case study of an internal platform, S Group, we show how the management of the cooperative was able to create the building blocks through an organizational transformation that served as the foundation for an efficient internal platform. By 1996, S Group was a platform where participants actively cooperated with each other, and after decades of suboptimization, it was able to find efficiency through its platform structure and eventually become the market leader in the Finnish grocery retail market by the mid-2000s. Overall, the S Group internal platform is not a specific organizational structure similar to the definition by Ciborra (1996) but more a virtual organizing structure embedded across the organization. This virtual structure enabled decentralized parts to experiment and share new initiatives with each other and the headquarters to expand successful initiatives often leading to unconventional results similar to the case of Olivetti (e.g., Ciborra 1996). There are also repercussions for management. For example, in an internal platform like S Group, the central organization had to constantly shuffle between the group-level interests and the interests of each individual cooperative. This makes managing such an organization difficult.

Through the change process that took place between 1983 and 1996, the management of the cooperative was able to make the most use of the platform-like structure by creating three building blocks, (1) the divestment of value-destroying and non-valuable resources, (2) preventing exploitative resource usage, and (3) enabling participants to identify and create new activities, which created the foundation for efficiency and enabled the future success of the organization.

First, the new management started the process of divesting value-destroying and non-valuable resources across the cooperative. We refer to resources as the firm resources, including all the assets, capabilities, organizational processes, firm attributes, information, and knowledge that improve its efficiency and effectiveness (e.g., Barney 1991). By choosing to focus on the grocery retail sector as the flagship industry for S Group, the organization was able to slowly get rid of the diversified businesses it owned in order to develop a clear strategic direction for the future. The layoffs and forced mergers of cooperatives in the 1980s helped streamline operations, and although the process was slow, it allowed change to concretely come to the organization. Coming into the 1990s, S Group had a clear business portfolio in grocery retail through its national retail chains. SOK created the chain concepts, and the cooperatives ran the chains according to standardized principles with some modification to adjust for regional customer preferences.

Secondly, SOK created and enforced new rules and practices to address the problem of suboptimization. The successful information gathering in the early 1980s had allowed SOK to finally understand the weak financial position of S Group, and the necessary changes were put in place fairly quickly by the new management. Both the launch of Inex Partners and the new rules that made the cooperatives profit centers in S Group resulted in a mind-set change. The increased power of the SOK board and the reshuffling of cooperative CEOs resulted in each regional cooperative following the agreed strategy. This was crucial as although each cooperative was autonomous, by the end of the 1980s, the exploitative resource usage had stopped and decisions were made on a group level.

Thirdly, SOK management started the process of identifying and creating new activities across the S Group. By activities we mean any action undertaken by S Group's employees/stakeholders for the purpose of generating profits or developing economic opportunities. Previously there had been little coordination between what was done at the central organization and in the cooperatives. By the late 1980s, S Group was finally systematically developing new concepts such as hypermarkets and the bonus system. Although both were not entirely new concepts as they had been piloted before, the previously dysfunctional structure meant that information about these trials had not flown across the organization and the potential benefits of these activities were thus not realized.

Through these three building blocks, S Group was able to find synergies and make the most use of the platform-like structure in the organization. S Group was able to find new competitive advantage through the stronger cooperation that now took place between the central organization and the independent cooperatives. Several new business initiatives were launched in the 1990s which paved way for the future success of the organization. For example, as a result of the synergies provided by a more efficient organization altogether, one significant part of S Group's new competitive advantage in the 1990s was in logistics and purchasing where it had been the most behind its competitors only a decade earlier.

Conclusion

In this chapter, we contribute to platform literature by providing an empirical example of how organizational networks function as internal platforms. The single case study shows that the inability to effectively maximize the benefit of a platform-like organizational structure can cause significant problems and suboptimization for an organization. There is thus a lot that managers can learn from platform theory when trying to optimize the performance of decentralized organizations. Through the steps taken by the management throughout the 1980s and 1990s, our single case organization S Group was able to shift away from decades of suboptimization toward becoming a more functional organization that made the most use of its platform-like structure. Through this transformation, S Group was able to find new competitive advantage from several strategic initiatives launched at the time which helped pave way for the organization's future success as partly due to these changes by the end of the 2000s, it had become the market leader of the Finnish grocery retail industry.

So far, research on internal platforms has focused almost exclusively on product platforms and empirical examples from new product development. This chapter however is the first to expand this stream of platform literature outside the product development and engineering context, incorporating platform theory to study an organizational network. Based on our single case study, we suggest a few topics for further research in this domain. First, future research could examine how platform-like organizations in different contexts evolve over time. Our study provides one example of how an organizational transformation including the divestment of unprofitable businesses and structural change was needed in order to maximize the efficiency of the platform-like structure. Secondly, future research could focus on the relationships and dynamics of internal platform ecosystems such as our case study organization. Our study provides an example of how the cooperative became efficient once the headquarters had taken a more active role as only a support function for the independent cooperatives (or platform participants), for example, after it no longer was allowed to make a profit of its own. Finally, future studies could focus on better incorporating the logic of platforms to the management of decentralized organizations like organizational networks. We suggest that efficient organizational networks require a culture where each stakeholder including middle-management can propose, plan and initiate new strategic initiatives, similar to the logic of value creation in platform ecosystems that exist in platform literature. Our study provides an example of how the lack of this kind of culture had resulted in suboptimization as each independent cooperative aimed to maximize the value of their own business rather than contribute to maximizing the value of the whole organization.

Although as a single case study this chapter has its limitations in terms of generalizability, we draw some reasonable managerial implications from the study. A platform-like organization, such as the single case study we have introduced in this chapter, requires new capabilities from managers as they need to constantly juggle between the interests of the individual platform participants and the platform as a whole. In our single case study, a mind-set shift was required so that it was possible to make decisions that maximized the value of the total organization, even if one part may have suffered as a result. As by definition, a platform value is generated through the sum of interactions and exchanges taking place in the ecosystem, and the management needs to make sure that the platform's participants do not suboptimize or exploit other parts of the platform. In a decentralized organization such as in the single case study presented in this chapter, this is easier said than done.

References

Adner, R., & Kapoor, R. (2010). Value creation in innovation ecosystems: How the structure of technological interdependence affects firm performance in new technology generations. *Strategic Management Journal, 31*(3), 306–333.

Armstrong, M. (2006). Competition in two-sided markets. *The RAND Journal of Economics, 37*(3), 668–691.

Autio, E., & Thomas, L. (2014). Innovation ecosystems. In *The Oxford handbook of innovation management* (pp. 204–288). Oxford: Oxford University Press.

Baldwin, C. Y., & Woodard, C. J. (2008). *The architecture of platforms: A unified view* (Harvard Business School Finance Working Paper, 09-034).

Barney, J. (1991). Firm resources and sustained competitive advantage. *Journal of Management, 17*(1), 99–120.

Burgelman, R. A. (2011). Bridging history and reductionism: A key role for longitudinal qualitative research. *Journal of International Business Studies, 42*(5), 591–601.

Ciborra, C. U. (1996). The platform organization: Recombining strategies, structures, and surprises. *Organization Science, 7*(2), 103–118.

Corley, K. G., & Gioia, D. A. (2004). Identity ambiguity and change in the wake of a corporate spin-off. *Administrative Science Quarterly, 49*(2), 173–208.

Cusumano, M. A., & Gawer, A. (2002). The elements of platform leadership. *MIT Sloan Management Review, 43*(3), 51.

Eisenmann, T. R. (2006). *Winner-take-all in networked markets* (Harvard Business School Technical Note, 806-131).

Eisenmann, T., Parker, G., & Van Alstyne, M. (2011). Platform envelopment. *Strategic Management Journal, 32*(12), 1270–1285.

Eriksson, P., & Kovalainen, A. (2008). *Qualitative research in business studies*. London: Sage.

Flyvbjerg, B. (2006). Five misunderstandings about case-study research. *Qualitative Inquiry, 12*(2), 219–245.

Gawer, A. (2014). Bridging differing perspectives on technological platforms: Toward an integrative framework. *Research Policy, 43*(7), 1239–1249.

Gawer, A., & Cusumano, M. A. (2002). *Platform leadership: How Intel, Microsoft, and Cisco Drive industry innovation*. Boston: Harvard Business School Press.

Gawer, A., & Cusumano, M. A. (2014). Industry platforms and ecosystem innovation. *Journal of Product Innovation Management, 31*(3), 417–433.

Gioia, D. A., Corley, K. G., & Hamilton, A. L. (2013). Seeking qualitative rigor in inductive research: Notes on the Gioia methodology. *Organizational Research Methods, 16*(1), 15–31.

Hagiu, A., & Wright, J. (2015). Multi-sided platforms. *International Journal of Industrial Organization, 43*, 162–174.

Haucap, J., & Heimeshoff, U. (2014). Google, Facebook, Amazon, eBay: Is the internet driving competition or market monopolization? *International Economics and Economic Policy, 11*(1–2), 49–61.

Jacobides, M. G., Knudsen, T., & Augier, M. (2006). Benefiting from innovation: Value creation, value appropriation and the role of industry architectures. *Research Policy, 35*(8), 1200–1221.

Langley, A. (1999). Strategies for theorizing from process data. *Academy of Management Review, 24*(4), 691–710.

Langley, A., Smallman, C., Tsoukas, H., & Van de Ven, A. H. (2013). Process studies of change in organization and management: Unveiling temporality, activity, and flow. *Academy of Management Journal, 56*(1), 1–13.

McGrath, M. E. (1995). *Product strategy for high-technology companies*. New York: Irwin Professional Publishing.

McIntyre, D. P., & Srinivasan, A. (2017). Networks, platforms, and strategy: Emerging views and next steps. *Strategic Management Journal, 38*(1), 141–160.

Meyer, M. H., & Lehnerd, A. P. (1997). *The power of product platforms: Building value and cost leadership*. New York: Free Press.

Provan, K. G., & Kenis, P. (2008). Modes of network governance: Structure, management, and effectiveness. *Journal of Public Administration Research and Theory, 18*(2), 229–252.

Robertson, D., & Ulrich, K. (1998). Planning for product platforms. *MIT Sloan Management Review, 39*(4), 19–31.

Rochet, J. C., & Tirole, J. (2003). Platform competition in two-sided markets. *Journal of the European Economic Association, 1*(4), 990–1029.

Rumelt, R. P., Schendel, D. E., & Teece, D. J. (1994). *Fundamental issues in strategy: A research agenda*. Boston: Harvard Business Press.

Sanderson, S., & Uzumeri, M. (1995). Managing product families: The case of the Sony Walkman. *Research Policy, 24*(5), 761–782.

Sanderson, S. W., & Uzumeri, M. (1996). *Managing product families*. New York: Irwin.

Siggelkow, N. (2007). Persuasion with case studies. *Academy of Management Journal, 50*(1), 20.

Simpson, T. W. (2004). Product platform design and customization: Status and promise. *Artificial Intelligence for Engineering Design, Analysis and Manufacturing, 18*(1), 3–20.

Suddaby, R. (2010). Editor's comments: Construct clarity in theories of management and organization. *Academy of Management Review, 35*(3), 346–357.

Thomas, L. D., Autio, E., & Gann, D. M. (2014). Architectural leverage: Putting platforms in context. *The Academy of Management Perspectives, 28*(2), 198–219.

Tiwana, A. (2014). *Platform ecosystems: Aligning architecture, governance, and strategy*. Burlington: Morgan Kaufmann.

Tsoukas, H., & Chia, R. (2002). On organizational becoming: Rethinking organizational change. *Organization Science, 13*(5), 567–582.

Van Alstyne, M. W., Parker, G. G., & Choudary, S. P. (2016). Pipelines, platforms, and the new rules of strategy. *Harvard Business Review, 94*(4), 54–60.

Yin, R. K. (1994). *Case study research – Design and methods* (2nd ed.). Thousand Oaks: Sage Publications.

Applying the Extended Theory of Planned Behavior to Predict Collaborative Consumption Intentions

Arto Lindblom and Taru Lindblom

Abstract In this study we test and validate the theory of planned behavior (TPB) to predict consumers' collaborative consumption intentions. In addition, we extend the TPB by studying consumers' price consciousness as a potential determinant of collaborative consumption intentions. The empirical data is based on the survey study that was conducted in Finland in 2015. Our main findings are as follows: first, our study indicates that consumers' attitudes toward collaborative consumption are positively related to their intentions to participate in collaborative consumption. Second, our study reveals that subjective norm is positively related to consumers' collaborative consumption intentions. Our results also demonstrate that if consumers have the abilities to engage in collaborative consumption, it enhances their intention to participate in such behavior. As a final contribution, our study indicates that price consciousness acts as determinant for taking part in collaborative consumption. Overall, our results are consistent with the TPB. Based on our study, it can be argued that TPB is a useful theoretical framework to investigate the motivations among consumers to engage in collaborative consumption.

Keywords Sharing economy · Collaborative consumption · Theory of planned behavior

A. Lindblom (✉)
Department of Marketing, Aalto University School of Business, Helsinki, Finland
e-mail: arto.lindblom@aalto.fi

T. Lindblom
Faculty of Social Sciences, Sociology, University of Tampere, Tampere, Finland
e-mail: taru.lindblom@uta.fi

© Springer Nature Singapore Pte Ltd. 2018

A. Smedlund et al. (eds.), *Collaborative Value Co-creation in the Platform Economy*, Translational Systems Sciences 11,
https://doi.org/10.1007/978-981-10-8956-5_9

Introduction and Purpose

The "sharing economy" is spreading rapidly and has transformed people's ways of thinking about ownership and consuming. Matzler et al. (2015, p. 71) highlight this change in people's minds and behavior, stating that "While individuals have traditionally often seen ownership as the most desirable way to have access to products, increasing numbers of consumers are paying to temporarily access or share products and services rather than buy or own them." In a similar vein, Bardhi and Eckhardt (2012) state that consumers now want access to products, and they prefer to pay for the experience of using the object than to buy and own it (see also Watkins et al. 2016; Lindblom and Lindblom 2017).

In short, the "sharing economy" is the idea of sharing of resources between individuals who have access to goods as needed. Belk (2007, p. 126) has defined sharing "as the act and process of distributing what is ours to others for their use and/or the act or process of receiving or taking something from others for our use." Bardhi and Eckhardt (2012) argue that sharing is a mode of behavior that does not involve a transfer of ownership. Anyone can share almost anything: material goods, time, ideas, skills, and competencies. In practice, sharing can include carpooling, space-sharing, or couch surfing.

Perhaps the best-known form of sharing is collaborative consumption. Botsman (2013) defines collaborative consumption as "an economic model that is based on sharing, swapping, trading, or renting products and services" (see also Botsman and Rogers 2010). Belk (2014) makes an important distinction between collaborative consumption and other forms of sharing by defining collaborative consumption as a behavior where people coordinate the acquisition and distribution of resources for a fee or other compensation. One practical example of collaborative consumption is peer-to-peer renting (see, e.g., Philip et al. 2015).

Nowadays, there are several business ventures and startups that have boosted collaborative consumption to rapid growth by creating online-based platforms or marketplaces. Well-known examples of such ventures include Airbnb, an online accommodation marketplace, and Uber, a transportation network company. There are also many C2C platforms such as Mercari and Rakuma that focus on the collaborative consumption of second-hand or vintage goods. These new business models have gained a lot of interest among researchers and practitioners as well (see, e.g., Van Alstyne et al. 2016).

A growing body of research has extended our knowledge of the popular movement of collaborative consumption (see, e.g., Hamari et al. 2016). Researchers have tried to specify the reasons for participation in collaborative consumption and how collaborative consumption varies across various groups (see, e.g., Möhlmann 2015; McArthur 2015; Philip et al. 2015). For example, Möhlmann (2015) showed that consumers engage in collaborative consumption for rational reasons. In their report, Owyang et al. (2014) indicated that younger people are much more likely to engage in collaborative consumption than older people. In addition, Owyang et al. (2014)

found a small gender gap in collaborative consumption in that women are slightly more likely than men to engage in collaborative consumption.

In order to expand the collaborative consumption to a broader mass of consumers and to develop new kinds of commercial platforms, researchers and practitioners require more knowledge and understanding about the motivating factors to engage in collaborative consumption (see, e.g., Akbar et al. 2016). The theory of planned behavior (TPB) offers an interesting framework to examine the factors that are related to consumers' intentions to engage in collaborative consumption. TPB proposes that consumers' intentions can best be predicted by their attitudes, subjective norms, and perceived behavioral control (Ajzen 1991, 2005). To our knowledge, no prior academic research has used TPB to investigate the determinants of collaborative consumption. Against this backdrop, we test and validate TPB to predict consumers' collaborative consumption intentions. In addition, we extend the TPB by studying consumers' price consciousness as a potential determinant of collaborative consumption intentions. Previous studies have indicated that this factor may have great impact on consumers' participation in collaborative consumption (e.g., Möhlmann 2015; Philip et al. 2015).

To be more specific, we focus on the following questions:

1. How are consumers' attitudes toward collaborative consumption related to their intentions to engage in collaborative consumption?
2. How are subjective norms related to consumers' intentions to engage in collaborative consumption?
3. How is consumers' perceived behavioral control related to their intentions to engage in collaborative consumption?
4. How is consumers' price consciousness related to their intentions to engage in collaborative consumption?
5. How do these potential determinants (i.e., attitudes, subjective norms, perceived behavioral control, and price consciousness) and collaborative consumption intentions vary among certain demographic groups?

By studying these questions, we contribute to the consumer research in general and to the sharing economy and collaborative consumption literature in particular. Also, we aim to provide new insights for practitioners to develop and promote their platforms or marketplaces in the area of sharing of economy.

Theoretical Background

The theory of planned behavior (TPB), developed by Ajzen and Fishbein (1980) and Ajzen (1991), has proved useful in understanding a wide variety of individual behaviors (see, e.g., Eddosary et al. 2015; Paul et al. 2016). TPB has been tested and validated in settings, including blood donation, healthy eating, green product consumption, digital piracy, and alcohol use (see, e.g., Armitage and Conner 2001; Netemeyer and Bearden 1992; Paul et al. 2016).

TPB is an extension of Fishbein's theory of reasoned action (TRA) (Fishbein and Ajzen 1975). According to Dutta and Singh (2014), both models were designed to explain the factors that influenced behavior. The difference between these two models is that TRA (Fishbein and Ajzen 1975) does not include perceived behavioral control and thus is not designed to predict behaviors that are outside an individual's volitional control (Hassan et al. 2016).

A key factor in TPB is the individual's intention to perform a given behavior. In short, intentions are assumed to capture the motivational factors that affect the behavior in question (Ajzen 1991). Ajzen (1991, p. 181) argues that intentions are "indications of how hard people are willing to try, of how much of an effort they are planning to exert, in order to perform the behavior." In a similar vein, Dutta and Singh (2014) define intentions as individual's conscious motivation to make an effort to engage in the specific behavior. Ajzen (1991, p. 181) states that "the stronger the intention to engage in a behavior, the more likely its performance." In other words, intention is a reliable predictor of actual behavior although the relationship between intention and behavior is not perfect (Ajzen 1991, 2005; Yunhi and Heesup 2010).

According to TPB, intention has three antecedents (Ajzen 1991; Hrubes and Ajzen 2001; see also Ajzen 2005):

- Attitude toward behavior: the degree to which individual has a favorable (or unfavorable) assessment or appraisal of the behavior in question
- Subjective norms: perceived social pressure to perform (or not to perform) the behavior in question
- Perceived behavioral control: perceived ease (or difficulty) of performing the behavior in question

In short, TPB suggests that individual's attitude toward the behavior in question, individual's perception of whether peer group or other significant others want him/her to perform the behavior, and individual's perceived ability to do so will predict his/her intention to undertake the behavior (Ajzen 1991, 2005).

Research Hypotheses

Attitude toward the behavior, the first determinant of intention, can be defined as an individual's favorable or unfavorable assessment of the behavior under consideration (Ajzen 1991, 2005; Lee et al. 2009). In other words, attitude refers to judgment on whether the behavior in question is good or bad and whether the individual wants to engage in such behavior (see, e.g., Dutta and Singh 2014). As Ajzen (1991) has stated, the more favorable the attitude with respect to the behavior under consideration, the stronger should be an individual's intention to perform that behavior. Therefore, we can assume that if consumers have a favorable perception of collaborative consumption, they probably have a high intention to engage in such behavior. In other words, based on TPB, we predict the following:

H1 Favorable attitude toward collaborative consumption is positively related to consumers' intention to engage in collaborative consumption.

The second determinant of intention in the TPB is subjective norm. In short, subjective norms refer to an individual's estimate of the social pressures on him/her to engage (or not) in the behavior under consideration (Ajzen 1991, 2005; Dutta and Singh 2014; Yunhi and Heesup 2010; Paul et al. 2016). According to Marta et al. (2014, p. 199), subjective norms are "a function of the person's beliefs regarding what each referent thinks he or she should do and the motivation to comply with these referents." Although perceived social pressure may significantly affect a consumer's intentions to undertake a specific behavior, it has been argued that the predictive power of subjective norms is sometimes low (Ajzen 1991; Lee et al. 2009; Paul et al. 2016). In this study, subjective norms are perceived social pressure to engage in collaborative consumption. Based on TPB, we predict the following:

H2 Subjective norms supporting collaborative consumption are positively related to consumers' intention to engage in collaborative consumption.

The third predictor of intention in the TPB is perceived behavioral control. In short, perceived behavioral control is "an individual's perception of the ease or difficulty of conducting the behavior" (Ajzen 1991, p. 183; see also 2005). Marta et al. (2014) argue that perceived behavioral control addresses people's perceptions of their ability to perform a given behavior. Ferdous (2010), in turn, states that perceived behavioral control reflects past experiences and anticipated obstacles. According to Hansen (2008), a person is more likely to carry out a behavior if he or she perceives that carrying out such behavior is easy. Paul et al. (2016) state that several studies have shown that perceived behavioral control is positively linked with intention in various contexts. In this study, perceived behavioral control refers to whether consumers have the abilities to engage in collaborative consumption. Based on TPB, we predict the following:

H3 Perceived behavioral control is positively related to consumers' intention to engage in collaborative consumption.

It is widely argued that for many consumers saving money is one of the key reasons to engage in collaborative consumption. For example, Möhlmann (2015) showed that consumers engage in collaborative consumption mainly for rational reasons. According to Möhlmann (2015), consumers pay attention to the fact that collaborative consumption is a way to help them to save money. Philip et al. (2015) also found that people engaged in peer-to-peer renting for economic pursuits, to maximize savings and earnings. Owyang et al. (2014) revealed that for many consumers the reason for collaborative consumption is price. Although there are also some contradictory results (Moeller and Wittkowski 2010), it can be assumed that consumers' price consciousness can be a significant antecedent of collaborative consumption. Price consciousness refers to the importance that consumers give to price when evaluating or purchasing products (Lichtenstein et al. 1988, 1993). To be more specific, we predict the following:

H4 Price consciousness is positively related to consumers' intention to engage in collaborative consumption.

Methodology

Sample

The sample was drawn from the adult population of Finland (mainland). The sample thus is nationally representative. A total of 3500 questionnaires and self-addressed prepaid envelopes were mailed to Finnish citizens. Of these, 976 (27.9%) usable responses were returned. Table 1 lists the demographic details of the respondents.

Measures

Collaborative Consumption Intentions

Intentions are consumers' aim to carry out and engage in collaborative consumption in the future (see, e.g., Ajzen 1991, 2005). Based on this definition, we developed two items to measure collaborative consumption intentions:

- I have an intention to increase collaborative consumption in the near future.
- Collaborative consumption will not be a central part of my consumption in the future (reverse).

Both items were measured on a 7-point scale, and these items exhibited a reliability of 0.673.

As the terminology of collaborative consumption is not very established in Finland, a short introductory text explaining to respondents what is referred to by the concept of collaborative consumption was provided in the questionnaire. Examples of such behavior and the marketplaces where they take place were provided. In the questionnaire we used a Finnish word that refers to peer-to-peer consumption or trade.

Table 1 Characteristics of the respondents

	% ($n = 976$)
Gender	
Female	59.9
Male	40.1
Age	
Under 26	12.7
26–35	12.4
36–45	12.8
46–55	14.0
56–65	22.4
Over 65	25.8
Employment status	
Employed	40.5
Unemployed	4.9
Entrepreneur	5.6
Student	8.9
Retired	35.8
Parental leave	1.8
Other	2.5
Education	
Primary school	16.2
Vocational school	36.7
Upper secondary school	13.4
University/polytechnic degree	29.1
Academic postgraduate degree	4.6
Income quartile	
1st (lowest) 2900 € or less	34.4
2nd 2901–3900 €	15.2
3rd 3901–5900 €	27.1
4th (highest) 5901 € or over	23.3

Collaborative Consumption Attitudes

Collaborative consumption attitudes are the extent to which someone perceives collaborative consumption as favorable or unfavorable (Ajzen 1991, 2005; Lee et al. 2009). Based on this definition, we developed two items to measure collaborative consumption attitudes:

- I have favorable attitude toward collaborative consumption.
- Collaborative consumption is a smart way to acquire and sell the goods.

Both items were measured on a 7-point scale and exhibited a reliability of 0.916.

Subjective Norms

Subjective norms are perceived social pressure to engage in collaborative consumption (Ajzen 1991, 2005). Based on this definition, we developed two items to measure subjective norm:

- My friends or family members have made me interested in collaborative consumption.
- Many of my close friends have engaged in collaborative consumption.

Both items were measured on a 7-point scale and exhibited a reliability of 0.736.

Perceived Behavioral Control

Perceived behavioral control is whether consumers have the abilities to engage in collaborative consumption (Ajzen 1991, 2005). Based on this definition, we developed two items to measure perceived behavioral control:

- Acquiring and selling the goods by engaging in collaborative consumption is simple.
- Collaborative consumption is uneasy way to acquire and sell the goods (reverse).

Both items were measured on a 7-point scale and exhibited a reliability of 0.713.

Price Consciousness

Price consciousness is the degree to which consumers focus on paying low prices (Lichtenstein et al. 1988; Lichtenstein et al. 1993). Based on current literature on price consciousness, we utilized the following items in measuring price consciousness (see Kopalle and Lindsey-Mullikin 2003; Kukar-Kinney et al. 2012):

- I check the prices even for inexpensive items before buying.
- Low price is an important consideration in my purchases.
- No matter what I buy, I shop around to get the lowest price.
- I am not willing to make extra effort to find lower prices (reverse).

All items were measured on a 7-point scale and exhibited a reliability of 0.690.

Gender had values (1) female and (2) male.

Age was measured as continuous variable (year of birth) that was categorized to six age groups: (a) under 26, (b), 26–35, (c) 36–45, (d) 46–55, (e) 56–65, and (f) over 65. It should be noted that there was a substantial share of missing values for age (13.4%).

Employment status had seven categories: (a) employed, (b) unemployed or seeking for a job, (c) entrepreneur, partner in a company, or freelancer, (d) student, (e) retired, and (f) other.

Education indicated respondent's highest level of education and had the following categories: (a) primary/comprehensive school, (b) vocational school, (c) matriculation (upper secondary school), (d) university/polytechnic, (e) academic postgraduate diploma.

Income was measured as total monthly gross income of the household. The initial categorization had 10 categories at roughly 1000 euro intervals (the median income category was 3901–4900€), but for the analyses, the income brackets were recoded to approximate income quartiles. The income quartile categories are thus (a) 2900 euros or less, (b) 2901–3900 euros, (c) 3901–5900 euros, and (d) over 5900 euros. Since the original measure was categorized, the quartiles produced are approximate.

Data Analysis

Data analysis is executed in two parts. The first part of the data analysis utilizes analysis of variance (ANOVA) to test the differences between certain demographic groups across collaborative consumption attention (CCA), subjective norm (SN), perceived behavioral control (PBC), price consciousness (PC), and collaborative consumption intentions (CCI). The second part presents regression-based prediction models using CCA, SN, PBC, and PC as independent variables and CCI as the dependent variable.

The first stage of the data analysis used the ANOVA technique to compare the CCA, SN, PBC, PC, and CCI across several demographic groups. In Table 2, we present the means, F-values, and p-values produced by the ANOVA procedure.

As can be seen in Table 2, Finnish consumers have very positive attitudes toward collaborative consumption. On the average the respondents exhibited high interest in collaborative consumption in the form of attitudes (mean 5.36). The intentions to participate in collaborative consumption were not as high, though: mean being somewhat lower at 4.08. The means for other constructs were rather similar with CCI; means hovered around 4, indicating a rather neutral perception on SN, PBC, and PC.

The impact of the background variables varied greatly across the studied constructs. Age had clearly strongest effect in most cases. Collaborative consumption attitude was clearly highest in the age bracket 26–35. This age group had also highest scores for subjective norm. Perceived behavioral control was even across the age groups except for those over 65 who had significantly lower scores than the others. PC was highest among the youngest cohort and lowest among 36- to 45-year-olds. Collaborative consumption intention was highest for those less than 45 years old.

From a gender perspective, female consumers' intentions were more collaborative, both in terms of their attitudes and their intentions. Women also had slightly higher scores for subjective norm, whereas there was no statistical significance between the genders for their perception of behavioral control or price consciousness.

Table 2 Analysis of variance (ANOVA) regarding collaborative consumption attitude (CCA), subjective norm (SN), perceived behavioral control (PBC), price consciousness (PC), and collaborative consumption intention (CCI). Means, F-values, and significance (p)

	CCA		SN		PBC		PC		CCI	
Total	**5.36**		**4.44**		**4.29**		**4.16**		**4.08**	
		F (sig.)		F (sig.)		F (sig.)		F (sig.)		F (sig.)
Age		**17.86*** **		**17.46*** **		**5.85*** **		**9.45*** **		**19.72*** **
Under 26	5.71		4.79		4.20		4.83		4.38	
26–35	6.00		5.21		4.61		4.28		5.01	
36–45	5.86		4.85		4.60		3.90		4.52	
46–55	5.47		4.46		4.28		4.00		4.18	
56–65	5.34		4.42		4.52		4.06		4.05	
Over 65	4.69		3.82		3.95		4.11		3.37	
Gender		**5.14***		**28.46*** **		**1.13 ns**		**2.3 ns**		**7.49** **
Female	5.45		4.65		(4.26)		(4.21)		4.20	
Male	5.23		4.14		(4.35)		(4.09)		3.91	
Employment		**14.14*** **		**12.01*** **		**4.26*** **		**9.65*** **		**17.63*** **
Employed	5.66		4.71		4.48		4.05		4.45	
Unemployed	5.38		4.31		4.44		4.59		4.24	
Entrepreneur	5.54		4.56		4.52		3.45		3.73	
Student	5.81		4.89		4.41		4.77		4.46	
Parental leave	6.38		5.62		4.62		4.60		5.91	
Retired	4.82		3.96		4.01		4.15		3.50	
"Other"	5.58		4.37		4.06		4.35		4.04	
Education		**10.69*** **		**7.78*** **		**0.74 ns**		**4.04** **		**4.36** **
Primary	4.90		4.12		(4.22)		4.35		3.78	
Vocational	5.19		4.35		(4.31)		4.13		4.06	

Upper secondary	5.53		4.62		(4.31)		4.29		4.01	
"University"	5.76		4.76		(4.36)		4.14		4.37	
Postgraduate	5.24		3.87		(4.01)		3.59		3.73	
Income quartile		**3.49***		**0.75 ns**		**1.26 ns**		**14.65*****		**(0.60)ns**
1st (low)	5.26		(4.41)		(4.29)		4.48		(4.05)	
2nd	5.17		(4.37)		(4.31)		4.03		(4.06)	
3rd	5.40		(4.41)		(4.21)		4.12		(4.06)	
4th (high)	5.61		(4.57)		(4.46)		3.84		(4.22)	

Note: The scales for CCA, SN, PBC, PC, and CCI range from 1 to 7
When there was no or very little statistical significance, the means are presented in parenthesis
This category includes those having either university or polytechnic degree
Ns not significant
$*p < 0.05$; $**p < 0.01$; $***p < 0.001$; $^p = 0.05–0.10$

Table 3 Regression analysis on collaborative consumption intention (CCI), collaborative consumption attitude (CCA), subjective norm (SN), perceived behavioral control (PBC), price consciousness (PC), standardized beta coefficients, F-value, sig

	CCI unadjusted main effects	CCI model 1	CCI model 2	CCI model 3
CCA	0.526	0.307	0.348	0.518
	(F = 365.49***)			
	(R = 27.6)			
SN	0.549	0.367		
	(F = 413.53***)			
	(R = 30.1)			
PBC	0.549		0.386	
	(F = 410.16***)			
	(R = 30.1)			
PC	0.172			0.107
	(F = 28.44***)			
	(R = 2.8)			
$R^2$100*		36.2	39.3	29.1
F (Sig.)		270.99***	307.06***	190.66***

Employment status had a significant effect on the studied constructs. Respondents on parental leave had most positive attitudes and intentions toward collaborative consumption and highest scores for SN and PBC, whereas the retired proved to be their complete opposites. The CCA of entrepreneurs and unemployed were similar, although the intentions to participate in collaborative consumption set them apart (entrepreneurs being less eager). PC was lowest among the entrepreneurs and highest among students, the unemployed, and those on parental leave. This is quite likely an effect of income.

Education had an effect to all but one construct (PBC). The effect size of education was rather modest, being strongest for CCA. CCA was clearly lowest among those with only basic education and highest for university graduates. Interestingly enough the former and those with postgraduate degree also had the lowest scores for CCI. The holders of postgraduate degrees had the lowest scores for SN and PC across the education groups. The most PC groups were those with basic education and those with upper secondary degrees.

Predictably, income had the greatest impact on price consciousness. Income showed no statistical significance for SN, PBC, and CCA. A moderate effect was found between CCI and income: the higher income groups had more positive attitudes toward collaborative consumption.

The second stage proceeded with the analysis of the linear association by using regression analysis. Regression equations were developed to determine the relationship between the respondents' intentions toward collaborative consumption (CCI) and CCA, SN, PBC, and PC. Table 3 shows the equations presenting regression coefficients, F-values, significance (p), and coefficient of determination (adjusted R2).

There are three conclusions to be derived from Table 3. First, all the studied constructs but price consciousness have a substantial impact on CCI. This indicates that contrary to the previous assumptions in the literature, lower price is not a major motive for taking part in collaborative consumption. Second, each studied construct has a positive (linear) relationship with the CCI. Increase in scores for CCA, SN, PBC, and PC increase the values of CCI. Third, the constructs CCA, SN, and PBC each have a rather strong independent effect on CCI. This can be seen in models 1–3, where the effect of each construct remains even when controlling for another variable.

All constructs with the exception of PC explain a substantial share of the variation in CCI (R2 coefficients ranging from 27.6 to 30.1). As there is only a minor increase of R2 from the baseline model (unadjusted effects) to models 1–3, the CCI can be well explained by each individual construct. However, the CCI can be assumed to be highest among those who perceive their behavioral control to be high and who have positive attitudes toward collaborative consumption (as shown by model 2 in Table 3).

Thus, all four hypotheses were supported. First, support for H1 was found as CCA was positively related to CCI (standardized beta coefficient was positive at 0.53). Second, SN was positively related to consumer's intention to participate in collaborative consumption (standardized beta 0.55), thereby supporting H2. Third, there was a statistically significant association between PBC and CCI (standardized beta 0.55), so H3 was supported. Finally, PC was positively related to CCI (standardized beta 0.17) although the effect remained rather weak. Therefore we can also say H4 is confirmed but with some caution. In addition we found that each of these hypotheses work well alone, as there was no significant change in the models where several constructs were introduced. It means they can predict collaborative consumption intentions very well independently.

Conclusion and Discussion

In the past few years, there have emerged several studies that have extended our knowledge of the sharing economy and growing movement of collaborative consumption (see, e.g., Belk 2014; McArthur 2015; Möhlmann 2015; Lindblom and Lindblom 2017). In particular, various sharing platforms and marketplaces have gained a lot of interest among researchers and practitioners (see, e.g., Van Alstyne et al. 2016). Although there is an increasing interest toward sharing economy and collaborative consumption, there is great need for further research. Especially, there is a lack of understanding of the factors that affect consumers' intentions to engage in collaborative consumption. Against this background, our aim was to further increase understanding of determinants of collaborative consumption intentions. In addition, we analyzed how these determinants and intentions vary among certain demographic groups.

Based on our analyses and the sample of 976 Finns, our study offers the following substantive contributions:

As a first contribution, our study indicates that consumers' attitudes toward collaborative consumption are positively related to their intentions to participate in collaborative consumption. In other words, consumers who have favorable perception of collaborative consumption also have strong intentions to engage in collaborative consumption. While this result is more or less intuitive, it is well in line with the previous research that attitudes are one of the most important predictors of behavioral intentions (Ajzen 1991, 2005; see also Eddosary et al. 2015; Paul et al. 2016; Stanislawski et al. 2013).

As a second contribution, we demonstrated that subjective norm is positively related to consumers' intentions to engage in collaborative consumption. In other words, consumers are willing to perform collaborative consumption if they perceive that their significant others prefer this kind of behavior. This is a theoretically interesting finding because many of the previous studies have indicated that subjective norm has a very weak or even an insignificant link to intention (Ajzen 1991; Stanislawski et al. 2013; Paul et al. 2016). However, it seems that in the context of collaborative consumption, subjective norm is an important determinant of consumers' intentions.

As a third contribution, we confirmed that there is a straightforward relationship between perceived behavioral control and intentions. Our results demonstrated that if consumers have the ability to engage in collaborative consumption, it enhances their intention to participate in such behavior. Therefore, one could argue that in encouraging consumers to engage in collaborative consumption, it is important to enhance consumers' perceived behavioral control. By doing this, consumers' willingness to join in collaborative consumption could increase significantly. At the same time, it can be assumed that digitalization has already removed many obstacles to engaging in collaborative consumption.

As a final contribution, our study demonstrated that price consciousness is a determinant of collaborative consumption intentions. However, this link was relatively weak, and it might indicate that lower price may not be the key motive for taking part in collaborative consumption.

We also looked at how the determinants and intentions vary among demographic groups. Generally speaking, it seems that young and highly educated female consumers are most inclined to engage in collaborative consumption; elderly males with basic education and low income are less interested in collaborative consumption. These findings are well in line with results presented by Owyang et al. (2014).

For practitioners the results in this study have revealed interesting insights into Finnish consumers from collaborative consumption perspective. In particular, this study has increased the understanding of the determinants of collaborative consumption intentions. In short, these results could be useful for the companies that are planning to engage in online-based platforms or marketplaces in the area of sharing of economy.

Limitations and Future Research

There are some limitations that suggest caution in assessing our findings.

First, our study considered collaborative consumption in general, and therefore, findings may be different if specific collaborative consumption behaviors or contexts are considered. Future research should test and validate TPB in a variety of settings, such as peer-to-peer renting or carpooling.

Second, this study used a cross-sectional design. Therefore, it is difficult to establish causality between studied factors. In fact, in cross-sectional analyses, causality is often open to debate. However, to overcome this problem, future studies should employ longitudinal data to establish causality.

In addition, future studies should use more advanced analysis techniques such as structural equation modelling (SEM) to determine relationships between studied factors. Moreover, we argue that there is also a need for qualitative empirical studies to obtain a clearer understanding of collaborative consumption practices. These in-depth qualitative studies could reveal issues that would enable more thorough operationalization of the concepts linked to collaborative consumption.

References

Ajzen, I. (1991). The theory of planned behavior. *Organizational Behavior and Human Decision Processes, 50*, 179–211.

Ajzen, I. (2005). *Attitudes, personality, and behavior* (2nd ed.). Berkshire: Open University Press and McGraw-Hill.

Ajzen, I., & Fishbein, M. (1980). *Understanding attitudes and predicting social behavior*. Englewood Cliffs: Prentice Hall.

Akbar, P., Mai, R., & Hoffmann, S. (2016). When do materialistic consumers join commercial sharing systems. *Journal of Business Research, 69*(10), 4215–4224.

Armitage, C., & Conner, M. (2001). Social cognitive determinants of blood donation. *Journal of Applied Social Psychology, 31*, 1431–1457.

Bardhi, F., & Eckhardt, G. (2012). Access-based consumption: The case of car sharing. *Journal of Consumer Research, 39*, 881–898.

Belk, R. (2007). Why not share rather than own? *Annals of the American Academy of Political and Social Science, 611*, 126–140.

Belk, R. (2014). You are what you can access: Sharing and collaborative consumption online. *Journal of Business Research, 67*, 1595–1600.

Botsman, R. (2013, November 21). *The sharing economy lacks a shared definition*. Co.Exist. http://www.fastcoexist.com/3022028/the-sharing-economy-lacks-a-shareddefinition#4

Botsman, R., & Rogers, R. (2010). *What's mine is yours—The rise of collaborative consumption*. New York: HarperCollins.

Dutta, K., & Singh, S. (2014). Applying the theory of planned behavior to understand Indian housewives' purchase behavior towards healthy food brands. *Journal of Brand Management, 11*, 7–28.

Eddosary, M., Yong, J., Sagas, M., & Hee, Y. (2015). Consumers' intention to attend soccer events: Application and extension of the theory of planned behavior. *Psychological Reports, 117*, 89–102.

Ferdous, A. (2010). Applying the theory of planned behavior to explain marketing managers' perspectives on sustainable marketing. *Journal of International Consumer Marketing, 22*, 313–325.

Fishbein, M., & Ajzen, I. (1975). *Belief, attitude, intention, and behavior: An introduction to theory and research*. Reading: Addison-Wesley.
Hamari, J., Mimmi Sjöklint, M., & Ukkonen, A. (2016). The sharing economy: Why people participate in collaborative consumption. *Journal of the Association for Information Science and Technology, 67*, 2047–2059.
Hansen, T. (2008). Consumer values, the theory of planned behaviour and online grocery shopping. *International Journal of Consumer Studies, 32*, 128–137.
Hassan, L., Shiu, E., & Parry, S. (2016). Addressing the cross-country applicability of the theory of planned behaviour (TPB): A structured review of multi-country TPB studies. *Journal of Consumer Behaviour, 15*, 72–86.
Hrubes, D., & Ajzen, I. (2001). Predicting hunting intentions and behavior: An application of the theory of planned behavior. *Leisure Sciences, 23*, 165–178.
Kopalle, P., & Lindsey-Mullikin, J. (2003). The impact of external reference price on consumer price expectations. *Journal of Retailing, 79*, 225–236.
Kukar-Kinney, M., Ridgway, N., & Monroe, K. (2012). The role of price in the behavior and purchase decisions of compulsive buyers. *Journal of Retailing, 88*, 63–71.
Lee, R., Murphy, J., & Swilley, E. (2009). The moderating influence of hedonic consumption in an extended theory of planned behavior. *Service Industries Journal, 29*, 539–555.
Lichtenstein, D., Bloch, P., & Black, W. (1988). Correlates of price acceptability. *Journal of Consumer Research, 15*, 243–252.
Lichtenstein, D., Bloch, P., Black, W., Ridgway, N., & Netermeyer, R. (1993). Price perceptions and consumer shopping behavior: A field study. *Journal of Marketing Research, 30*, 234–245.
Lindblom, A., & Lindblom, T. (2017). De-ownership orientation and collaborative consumption during turbulent economic times. *International Journal of Consumer Studies, 41*, 431–438.
Marta, E., Manzi, C., Pozzi, M., & Vignoles, V. (2014). Identity and the theory of planned behavior: Predicting maintenance of volunteering after three years. *Journal of Social Psychology, 154*, 198–207.
Matzler, K., Veider, V., & Kathan, W. (2015). Adapting to the sharing economy. *MIT Sloan Management Review, 56*, 71–77.
McArthur, E. (2015). Many-to-many exchange without money: Why people share their resources. *Consumption, Markets & Culture, 18*, 239–256.
Moeller, S., & Wittkowski, K. (2010). The burdens of ownership: Reasons for preferring renting. *Managing Service Quality, 20*, 176–191.
Möhlmann, M. (2015). Collaborative consumption: Determinants of satisfaction and the likelihood of using a sharing economy option again. *Journal of Consumer Behaviour, 14*, 193–207.
Netemeyer, R., & Bearden, W. (1992). A comparative analysis of two models of behavioral intention. *Journal of the Academy of Marketing Science, 20*, 49–59.
Owyang, J., Samuel, A., & Grenville, A. (2014). *Sharing is the new buying*. Available online: www.web-strategist.com. Accessed 03 Dec 2014.
Paul, j., Modi, A., & Patel, J. (2016). Predicting green product consumption using theory of planned behavior and reasoned action. *Journal of Retailing & Consumer Services, 29*, 123–134.
Philip, H., Ozanne, L., & Ballantine, P. (2015). Examining temporary disposition and acquisition in peer-to-peer renting. *Journal of Marketing Management, 31*, 1310–1332.
Stanislawski, S., Yasushi, S., & Shuji, O. (2013). Green consumption and the theory of planned behavior in the context of post-megaquake behaviors in Japan. *Advances in Consumer Research, 41*, 321–325.
Van Alstyne, M., Parker, G., & Choudary, S. (2016). Pipelines, platforms, and the new rules of strategy. *Harvard Business Review, 94*(4), 54.
Watkins, R., Denegri-Knott, J., & Molesworth, M. (2016). The relationship between ownership and possession: Observations from the context of digital virtual goods. *Journal of Marketing Management, 32*, 44–70.
Yunhi, K., & Heesup, H. (2010). Intention to pay conventional-hotel prices at a green hotel—A modification of the theory of planned behavior. *Journal of Sustainable Tourism, 18*, 997–1014.

Distributed Governance in Multi-sided Platforms: A Conceptual Framework from Case: Bitcoin

Juri Mattila and Timo Seppälä

Abstract Over the last decade, blockchain technology has facilitated a method by which a network of equipotent and equally privileged peers can jointly maintain and edit databases in an entirely decentralized manner, without any kind of an intermediary exhibiting unilateral control. As a consequence it has enabled the creation of a new type of multi-sided platform architecture with distributed governance. As the different platform provision functions are opened to free market competition rather than monopolized by a single entity, the monopoly-like pricing structure typical of platforms is overhauled. Instead, blockchain-enabled distributed platforms appear to share value more evenly between the all the different market sides connected to the platform. Our analysis reveals that blockchain technology adds new considerations to how multi-sided platform architectures should be perceived and analyzed.

Keywords Platforms · Multi-sided markets · Governance · Blockchain · Bitcoin · Cryptocurrency

Background Introduction

In recent years, blockchain technology[1] has facilitated a method by which a network of equipotent and equally privileged peers can jointly maintain and edit databases in an entirely decentralized manner, without any kind of an intermediary exhibiting

[1] For this chapter, we define blockchain technology as the cryptographically concatenated data structure and network architecture described by Nakamoto (2008) which entails a proof-of-work consensus protocol and employs cryptographic tokens of value, more commonly referred to as cryptocurrency.

J. Mattila (✉) · T. Seppälä
The Research Institute of the Finnish Economy (ETLA), Helsinki, Finland

Aalto University School of Science, Espoo, Finland
e-mail: juri.mattila@etla.fi; timo.seppala@etla.fi

© The Author(s) 2018

A. Smedlund et al. (eds.), *Collaborative Value Co-creation in the Platform Economy*, Translational Systems Sciences 11,
https://doi.org/10.1007/978-981-10-8956-5_10

unilateral control. As a consequence, blockchain technology has enabled the creation of a new type of platform architecture with distributed governance. Rather than a single intermediary constructing and governing the platform, these blockchain-enabled distributed platforms utilize internal joint revenue models to incentivize open participation in all the different platform provision functions. This has enabled the provision of similar digital platform service systems as with a single platform provider but in a completely distributed manner (see Gawer and Cusumano 2002; Boudreau and Hagiu 2008; Gawer 2009; Hagiu 2014; Hagiu and Wright 2015; van Alstyne et al. 2016, Parker et al. 2016).

As the platform provision functions are opened to free market competition rather than monopolized by a single entity, the monopoly-like pricing structure typical of platforms where the platform provider captures most of the value is overhauled. Instead, blockchain-enabled distributed platforms appear to share value more evenly between all the different market sides connected to the platform: application developers, users, miners, nodes, and platform developers alike.

A considerable amount of literature has been published on platforms. These studies offer various definitions for a platform.[2] In the first stage of development, the terms "platform" and "product platform" were used by R&D scientists to illustrate the creation of new-generation products and services or a new product family for use as a basis for a range of product and service variants offered to customers in one-sided and two-sided markets (Gawer and Cusumano 2002; Eisenmann et al. 2006; Gawer 2009; van Alstyne et al. 2016).

In the second stage of development, a school of technology researchers defined a platform as a control point and as a gatekeeper role in industrial networks used for earning revenues without actually creating value while at the same time economically damaging the network as a whole (for control point discussion, see Cusumano and Selby 1995; Cusumano and Yoffie 1998; Gawer 2009; van Alstyne et al. 2016; for gatekeeper role discussion, see Ballon 2009a, b; Ballon and Van Heesvelde 2011; Pon et al. 2014).

In the third stage of development, industrial economists defined the term "platform" as a medium for conveying products, services, and related transactions—as a marketplace between two or more different market sides (Rochet and Tirole 2003, 2006; Parker and van Alstyne 2005; Boudreau and Hagiu 2008; Rysman 2009; Hagiu and Wright 2015; van Alstyne et al. 2016).

Numerous definitions have been proposed for multi-sided markets over the years. One of the earliest definitions revolved around the presence of indirect network effects between two or more groups of platform participants (Katz and Shapiro 1994; Caillaud and Jullien 2003; Armstrong 2006). Another approach has been to

[2] In this chapter, we use the term platform in reference to a system that allows the various actors—users, providers, and other stakeholders across organizational boundaries—to engage in value-adding activities. It is typical of platforms in this sense that the individual actors create, offer, and maintain mutually complementary products and services for various distribution channels and multi-sided markets within the framework of common business and contract rules, governance, and user experiences. A typical characteristic of a platform is to attract and lock in a range of actors keen to harness the economic benefits offered by direct and indirect network effects.

define two-sided markets through their pricing structure. According to this view, in a multi-sided market, profits are not only affected by prices—they are also essentially affected by how the prices are allocated to different participant groups (Rochet and Tirole 2003, 2006; Parker and van Alstyne 2005; Boudreau and Hagiu 2008; Rysman 2009).

Later on, Hagiu and Wright (2015) argued that a multi-sided platform is simply one which enables direct interactions between two or more market sides, each of which is somehow associated with the platform. Thus, a multi-sided platform typically comprises three kinds of elements: (1) a stable core, (2) a dynamic set of complementary assets, and (3) the design rules acting as interfaces between them. The generic idea behind platforms is that by facilitating the integration of various stable and dynamic elements in a manner carefully coordinated by the design rules, platforms can achieve a higher degree of innovative dexterity in some areas of interest while still preserving economies of scale in others (Teece 1986; Baldwin and Woodard 2009; Boudreau and Hagiu 2009; Bourdeau 2010; Parker et al. 2016).

Market structures can also be described as fixed-role and switch-role markets (Aspers 2008). Subsequently, the concept of a platform was extended to include social, primarily contractual boundary resources as well as technical boundary resources (Gawer 2009; Eisemann 2008; Bourdeau 2010; Yoo et al. 2010; Ghazawneh and Henfridsson 2013). Additionally, reference should be made to the research on technology platforms initiated in the 1990s as well as compatible and competing standards (Cusumano and Selby 1995; Kim and Kogut 1996; Cusumano and Yoffie 1998) and to the research on platform governance (Schilling 2005; Tiwana et al. 2010; Tiwana 2014).

Schilling (2005) defines platform governance as "the mechanisms through which a platform owner exerts influence over app developers participating in a platform's ecosystem." Boudreau and Hagiu (2009) analogize platforms to public regulators, with the exceptions that platforms typically regulate with an interest of maximizing profit by controlling prices, access, and interactions on the platform and that they usually exhibit regulatory behavior in the domain of technical design, system architecture, and technical relationships. Moreover, Parker and Van Alstyne (2014) define governance as the set of rules concerning who gets to participate in an ecosystem, how to divide the value, and how to resolve conflicts.

On a general level, the prevalent definitions of platform governance refer to the interaction between the platform provider and any agents who contribute to the service offering of the platform externally, from outside of the platform itself. While such *external governance* remains very relevant for research, in order to fully describe blockchain-enabled distributed platforms, the definition of platform governance must be expanded to include the mechanisms of *internal governance*; these are the mechanisms through which the providers of a multi-provider platform exert influence over other providers participating in the provision of the same platform (Schilling 2005; Tiwana et al. 2010; Tiwana 2014).

Parker et al. (2016) and de Reuver et al. (2017) propose that further studies on platform governance may yet provide answers to some of the open questions in platform literature, e.g., how to design balanced internal governance systems in multi-sided platforms to ensure fair operation. The internal governance structures in blockchain-enabled distributed platforms have not yet been thoroughly explored in the context of this open question in the multi-sided platform literature. By performing a case study on the most prominent blockchain-enabled distributed platform in existence—namely, the Bitcoin cryptocurrency network—this chapter makes an effort to determine whether blockchain architectures can constitute multi-sided platforms as delineated in the platform literature. The motivation for this analysis is to determine whether blockchain architectures are relevant to this open research question regarding more balanced internal governance systems in multi-sided platforms. Having established that these systems are indeed relevant, we then proceed to describe how the internal governance structure is organized in these new kinds of distributed platforms.

The aim of this study is to answer the following research questions:

1. To what extent is the concept of multi-sided platforms, as delineated in platform literature, compatible with blockchain architectures?
2. What kinds of platform governance structures do blockchain architectures exhibit when observed through this framework of multi-sided platforms?
3. What are the potential wider implications of blockchain governance structures on multi-sided platforms research?

Our analysis reveals that blockchain architectures add new considerations to how multi-sided platforms should be perceived. Firstly, in blockchain-enabled distributed platforms, not only have the product and service innovations pertaining to the platform been externalized, but the entire platform provision has been distributed across various market sides. Therefore, blockchain-enabled distributed platforms introduce new categories of interaction between platform participants. Secondly, blockchain-enabled distributed platforms introduce new models of platform governance. Furthermore, as no formal decision-making protocols are in place, an informal negotiation process takes place involving a scheme of theoretical attacks and countermeasures used as bargaining chips against other market sides. Thirdly, blockchain-enabled distributed platforms seem to have introduced a new method of platform monetization. Much like contemporary multi-sided platforms, blockchain-enabled distributed platforms also employ split revenue schemes but in a more equilateral manner and without monopolistic pricing structures.

The remainder of this chapter continues as follows. The second section explains the methodology of the study. The next section describes platform characteristics and the internal governance structure of blockchain-enabled distributed platform mechanism in the light of platform literature. The fourth section presents the findings of the research focusing on the following two key themes: (1) if blockchain-enabled distributed platforms meet the criteria of a multi-sided platform and (2) the applicability of Tiwana's (2014) framework on platform governance. We conclude this chapter with a short discussion on the implications for future research.

Research Methodology

The theoretical positioning of this research is mainly conceptual in nature. In academic literature, conceptual papers are ones quintessentially characterized by an integrational approach. Rather than emphasizing empirics, conceptual papers tend to fixate on providing cross-disciplinary insights and bridging theories. While the focus of conceptual research is typically much narrower in scope than that of theory-building research, a well-drafted conceptual paper may also contribute to theory by pointing out interesting relationships, improving theoretical coherence, and proposing new directions and perspectives for further research (Gilson and Goldberg 2015; Sutton and Staw 1995).

The chosen methodology for this study is a case study design. Case studies can be utilized for several different purposes, e.g., to test existing theories, to generate new ones, or simply to describe phenomena (Eisenhardt 1989). As its case of analysis, this paper examines the most prominent blockchain architecture in existence at this time: the Bitcoin cryptocurrency network. In the first part, we test the applicability of the multi-sided platform theory to blockchain architectures by superimposing the conceptual framework of multi-sided platforms onto the Bitcoin network. The second part of the analysis makes an effort to describe the phenomenon of internal and external governance in blockchain architectures from the perspective of multi-sided platforms. The governance mechanisms are delineated by using the classification of Tiwana (2014) on platform governance as a framework.

Tiwana (2014) distinguishes three dimensions of platform governance: decision rights, control, and pricing policies. Decision rights pertain to setting goals for what the participants should be able to achieve and how. These rights are divided into two dimensions, with two categories each: (1) platform decision rights vs. application decision rights and (2) strategic decision rights vs. implementation decision rights. Control is used to ensure that the behavior of all the different participants is in line with those goals. Tiwana (2014) lists four different categories of control: input control, output control, process control, and relational control. Pricing policies dictate how the value created by this aligned collaboration will be shared. They describe the incentives that are used to attract all the different markets sides and to encourage them to participate in the platform ecosystem.[3]

[3] It should be noted that some similar types of categorizations have been drafted before in literature regarding Bitcoin's governance. For example, Filippi and Loveluck (2016) divide the governance of Bitcoin into two categories: "governance *by* the infrastructure" (i.e., the Bitcoin protocol managing the community) and "governance *of* the infrastructure" (i.e., the community managing the protocol). From the multi-sided platforms' perspective, the former bears great similarity to the concept of boundary resources (see section "Boundary Resources") and pricing policies (see section "Pricing Policies"), while the latter rather resembles the concept of platform decision rights (see section "Decision Rights"). The categorizations are not directly interchangeable, however, as the control category in Tiwana's framework (see section "Control") seems to incorporate aspects of both of Filippi's and Loveluck's categories simultaneously. Respectively, Gasser et al. (2015), speaks of "the power of influencing the normative content of the rules" and "their social realization" in Bitcoin. In this vernacular, the first half, again, relates to Tiwana's concept of decision

In addition to literary references, the technical understanding on blockchain technology in this chapter also draws from nondirective interviews with industry experts from companies such as IBM, BitPay, Blockstream, Vaultoro, Colu, Bitreserve, Google, 21, Stellar, Monax, Ascribe, Prasos, and Fortum in 2015–2017.

Analysis

Platform Characteristics in Blockchain Networks

Network Effects

In economics, a network effect describes a situation in which the value that a user gets from using a system depends on how many other participants the system has. This dependency can be either *positive* or *negative*. Direct network effects occur when increased use of a product or a service benefits or harms the users of that particular product or service. In the online gaming community, for example, players experience direct network effects from additional players joining the gaming environment. Indirect network effects are in question when increased use of a product or a service benefits or harms the users of a different product or service. To follow with the earlier example, in the online gaming community, the players benefit indirectly from the presence of game developers participating in the community and vice versa (Katz and Shapiro 1994; Caillaud and Jullien 2003; Armstrong 2006).

In platform literature, network effects have been mentioned as one of the key characteristics of multi-sided platforms. Likewise, blockchain-based distributed platforms also live and die by network effects. In fact, the successful fostering of network effects is even more important in blockchain-enabled distributed platforms than in other platform types, because the security and the technical functionality of the platform are to a large extent based on them (see Nakamoto 2008).

The robustness of a blockchain network typically grows stronger with every additional miner and node, thus fortifying the network against malevolent attacks that, if successful, could ultimately render the entire platform useless (Nakamoto 2008; Laszka et al. 2015; Yli-Huumo et al. 2016). As this makes the platform more stable and safer to operate, the logical consequence is the attraction of more users to join the platform (see, e.g., Vasek et al. 2014). Respectively, increased user activity makes the platform financially more appealing to miners and application providers, as more paying customers translates into more opportunities for business and profit (Kroll et al. 2013; see also Alabi 2017). Similarly, a higher number of application providers can be postulated to draw more platform developers to the platform, as

rights, while the social realization somewhat equates to Tiwana's notion of control. However, in the context of platform literature, so far, there has been little in the way of detailed conceptualizations of blockchain governance mechanisms.

this translates to better funding opportunities (van Wirdum 2016). The addition of more developers, in turn, increases usability and security, thus likely attracting more users, miners, application providers, and so on (see, e.g., Alabi 2017).

Multi-sided Markets

In platform literature, one characterization of multi-sided platforms is that they operate in *two-sided*—or more generally speaking, *multi-sided markets*. A market is said to be multi-sided if more than one market side is crucial to the outcomes of interest and the market sides exhibit network-effect-like externalities between them (Rochet and Tirole 2003; Parker and van Alstyne 2005; Rochet and Tirole 2006; Eisenmann et al. 2006; Boudreau and Hagiu 2008; Rysman 2009; Hagiu 2014; Hagiu and Wright 2015; van Alstyne et al. 2016).

Blockchain-based distributed platforms emerge from the collaboration of several different types of actors. This collaboration can be described as a multi-sided market where some of the market sides are platform providers and some of them are platform users. In the Bitcoin platform, five market sides can be distinguished: the users, the application providers, the miners, the nodes, and the platform developers.

Users are the actors whose main motive for participating in the distributed platform is to make use of its practical functionality. Users may be looking to transfer funds over the Internet, for example. Alternatively, they may be interested in acquiring some of the cryptocurrency tokens native to the platform with the intent to hold them as an investment for financial gain. Users can access the platform either by going through the services of application providers or by setting up their own node (see below) and connecting it to the network (Athey et al. 2016).

Application providers participate in the distributed platform by building products and services on top of it which extend the functionality and the usability of the underlying platform. In doing so, the application providers introduce complementarities which further enhance the network effects of the platform. Most often the application providers monetize their business by charging service fees from the users. Wallet service providers and cryptocurrency exchange services, for example, fall into this category (Athey et al. 2016).

Miners are essential to the operation of the network. They handle the data entries and the transactions between the users of the platform and provide security by partaking in the consensus-forming process of the network. By constantly solving concatenated mathematical problems in a cryptographic process known as *hashing*, the miners produce *proof of work*: a testimony to the fact that the content of the distributed database is authentic. As the hashing process consumes CPU power, the miners are incentivized to partake in the process by issuing them new cryptocurrency as *mining rewards*, as well as *transaction fees* charged from the users (Böhme et al. 2015; Gasser et al. 2015; Filippi and Loveluck 2016; Dimitri 2017).

Nodes are the computers/software clients which form the actual blockchain network. Each one of them individually maintains the distributed database which underlies the distributed platform. The nodes are also in charge of enforcing the consensus rules of the network by validating and propagating the new blocks produced by the miners for the blockchain. Each node operates autonomously, irrespective of other nodes or platform participants. Nodes can either be *full nodes* or *simplified payment verification (SPV) nodes*. Each full node maintains a full copy of the entire blockchain, making them the backbone of trustless security in the network. SPV nodes only store block headers, making them less independent but much lighter to run (Nakamoto 2008; Cawrey 2014; Filippi and Loveluck 2016).

Platform developers work on the technical design of the Bitcoin protocol. They formulate and propose adjustments to the technical and social boundary resources that govern the interactions that take place on the platform. Early stage developers can monetize their development efforts by holding an initial investment of tokens within the platform. As the amount of activity on the platform increases, so does the demand for the tokens, consequently driving up their purchasing power. At later stages, platform developers can also be supported directly by consortiums of large-scale application developers (Filippi and Loveluck 2016; van Wirdum 2016).

It is noteworthy that the market sides presented here are not mutually exclusive. For example, miners and application providers may run nodes to increase the robustness of the network, and platform developers may double up as users in the investor role. Any individual can freely join any number of the market sides, as long as they adhere to the predetermined protocols. To express the matter in the terms of platform literature, the Bitcoin platform constitutes switch-role markets rather than fixed-role markets (Aspers 2008).[4]

Complementarities

In platform literature, the presence of complementarities has been considered one key characteristic of multi-sided platforms. Goods and services are said to be complementary to one another if the utility offered by one greatly depends on the consumption of the other. One classic example of a complementarity is the relationship between a rowing boat and a pair of oars: these assets offer much higher utility when used together compared to when using them separately (for more information on complementary assets, see the seminal work of Teece 1986, 1988; Yoffie and Kwak 2006; Dahlander and Wallin 2006; Gawer and Henderson 2007).

The complementarities in blockchain-enabled distributed platforms do not only manifest themselves in the external goods and services attached to the platform but also internally within the platform itself. As blockchain-enabled distributed

[4] It should also be acknowledged that many of the market sides specified in this paper contain several factions which are engaged in internal power struggles within the market sides. For example, developers have mostly organized themselves into rival developer communities, miners have diversified their risks by forming collective mining pools which compete against one another, and so on. (Böhme et al. 2015) However, the scope of this research does not permit us to delve deeper into these internal power struggles, as our main focus is on the power relations and the power mechanics between the different market sides.

platforms do not have a distinct platform owner, they require the collaboration of various market sides to produce all the necessary platform provision functions (Filippi and Loveluck 2016). This collaboration can only produce a functional distributed platform if all the required provision functions are addressed.

Boundary Resources

In platform literature, boundary resources are the operational regulations and technical tools and interfaces governing the interaction between the platform owner and the platform participants. They can be used either to encourage platform development or to restrict it in places where the platform owner wishes to maintain control over the developmental direction of the platform. These resources are sometimes divided into *technical* and *social* boundary resources (Gawer 2009; Bourdeau 2010; Yoo et al. 2010; Ghazawneh and Henfridsson 2013).

Technical boundary resources govern the technical interactions between platform participants. In distributed platforms they manifest themselves in the operational principles of the distributed consensus network which underlies the platform. These boundary resources define operational features, such as how the nodes of the network connect and communicate to one another, what kind of a consensus protocol is employed by the network, and what are the prerequisites for partaking in the platform provision functions (Nakamoto 2008).

Social boundary resources manifest themselves as the predetermined framework for social interaction on the platform, e.g., prespecified terms of agreement for application developers, or revenue split models between participants (Gawer 2009). In distributed platforms, the social boundary resources consist of the business and contract rules governing the content of the distributed database that the distributed consensus network maintains. These rules define what kinds of modifications can be done to the database, by whom and in what manner. For example, in cryptocurrency platforms, users are typically not allowed to make payments that would exceed their account balance (see, e.g., Filippi and Loveluck 2016).

The social boundary resources of blockchain-enabled distributed platforms also outline the joint monetization models, e.g., how the mining rewards are added to the total money supply of the platform. It is noteworthy, however, that in blockchain-enabled distributed platforms, these split-revenue schemes are algorithmically governed, rather than decided by a platform owner (Dimitri 2017; Böhme et al. 2015). It seems to be the case then that blockchain-enabled distributed platforms bring some of the boundary resources formerly considered more social in nature into a more technical domain.

Contemporary platforms are also differentiated from blockchain-enabled distributed platforms by the fact that whereas the former are provided by one party, distributed platforms, by definition, have multiple equipotent and equally privileged platform providers. Therefore, distributed platforms also have a completely new area where boundary resources apply: the internal interactions between one platform provider and another platform provider. These inter-provider boundary resources include many familiar ones, such as APIs, SDKs, and technical documentations, but also some completely new ones, such as a consensus protocols and the algorithmically defined game-theoretical incentivization structures mentioned above (Nakamoto 2008).

Mechanisms of Internal Governance

Decision Rights

Tiwana (2014) divides decision rights into two dimensions, with two categories each: (1) platform decision rights vs. application decision rights and (2) strategic decision rights vs. implementation decision rights. Platform decision rights pertain to decisions relating to the platform, whereas application decision rights pertain to decisions relating to the complementary assets of the platform. Strategic decision rights refer to the right to determine *what* the platform or a complementary asset should be able to achieve, while implementation decision rights are related to determining how those goals should be accomplished.

Tiwana (2014) measures these four categories of decision rights on a gradient scale from full centralization to full decentralization. Baran (1964), however, differentiates between three configurations for communications networks: centralized, decentralized, and distributed. In order to more accurately describe the allocation of decision rights in blockchain-enabled distributed platforms, this paper expands Tiwana's (2014) gradient scale accordingly to include these three configurations (Fig. 1).

The problem with trying to establish joint platform strategic decision-making processes for blockchain-enabled distributed platforms (e.g., democratic voting) is that there is no clear and objective way to measure the support for a strategy among all the different platform providers and participants. Moreover, no single faction or individual has the power to dictate platform strategy on their own without sufficient support from the others (Filippi and Loveluck 2016; Gasser et al. 2015; Kroll et al. 2013). Therefore, the different market sides must communicate with one another to

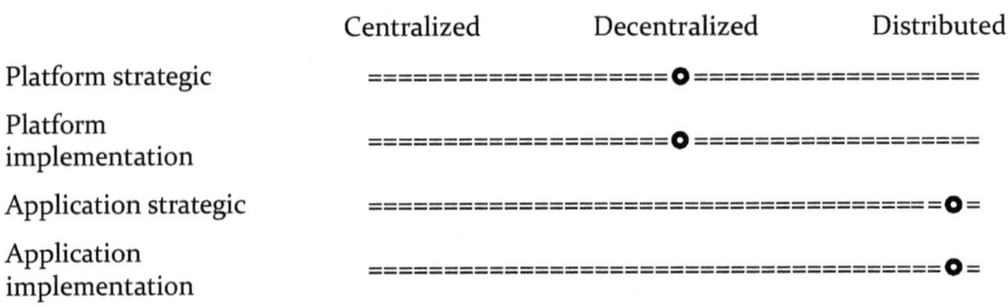

Fig. 1 Decision rights in Bitcoin

negotiate strategic platform decisions and their implementations, despite the fact that no structured forum or protocol exists for such negotiations at this point in time.

So, in theory, Bitcoin is anarchistic in governance, as no single platform owner is in control of the system and no formal mechanisms of multi-party decision-making are in place. A more detailed examination, however, reveals that some de facto structure exists regarding how platform decision rights are allocated within the system (Kroll et al. 2013; Gasser et al. 2015; Filippi and Loveluck 2016). For example, miners and nodes can signal their support by running different client versions, and developers can signal their support by committing to different development projects (Gasser et al. 2015; Srinivasan and Leland 2017).

It has been speculated by some that more formal protocols for strategic platform decision-making may be seen in the future. For example, anonymous cryptocurrency-based voting mechanisms may be considered for measuring how much support different planned strategies have in the platform ecosystem (Consensus 2017).

The platform implementation decisions in Bitcoin are decentralized rather than distributed in nature, as in practice the platform developer communities have great preeminence in what kinds of implementation proposals are brought forward in the platform ecosystem (Kroll et al. 2013; Gasser et al. 2015). It is mainly up to the miners and the nodes to decide which proposals are accepted—other parties' decision rights in this respect are mostly manifested in their ability to affect the decision-making of the aforementioned two market sides (Kroll et al. 2013; Atzori 2015; Filippi and Loveluck 2016).

As Bitcoin is based on a *permissionless* blockchain architecture, no permission is required from any party to become a part of the network (Filippi and Loveluck 2016). Therefore, application-related decision rights are completely distributed across the individual application providers and their respective developers.

Control

Output control, or *metrics*, refers to the mechanisms of rewarding or penalizing the platform participants on the basis of their performance against some pre-defined target performance metrics (Ouchi 1979; Tiwana 2014). In modern day platforms, explicit forms of output control are somewhat rare, as most platforms simply maintain one output criterion: the survival and success of their complementary assets in the free market competition environment (Armstrong 2006; Bester and Krähmer 2008).

In Bitcoin, output control is mainly exercised by users and platform developers (see Picture 1). While the platform developers cannot force anyone to run their software, they can make proposals for new source code implementations.

Blockchain-enabled distributed platforms rely on open source code. Therefore, in theory, anyone could write and implement their own version of the code, as long as it adheres to the consensus protocol of the network (Gasser et al. 2015; Filippi and Loveluck 2016). In practice, however, most if not all of the seriously considered suggestions for new source code implementations tend to come from the Bitcoin

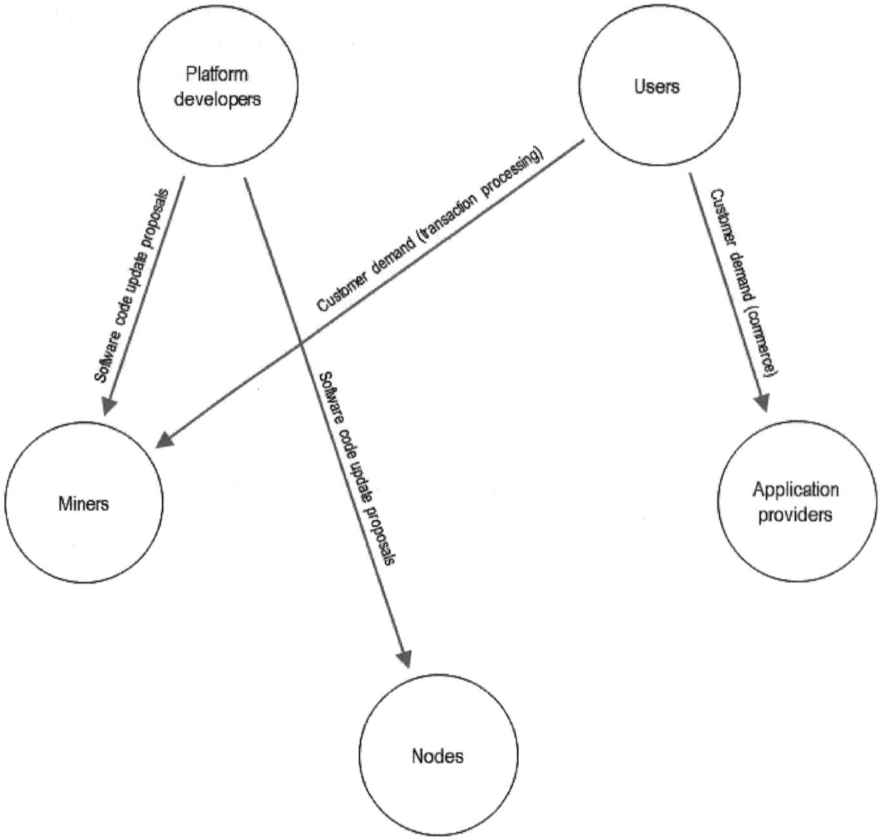

Picture 1 Output control in the Bitcoin platform

developer community. Therefore, the developer community has a great amount of control over what kinds of source code modifications are suggested to the network and what kinds of performance metrics and reward schemes they entail (Kroll et al. 2013; Atzori 2015; Gasser et al. 2015; Filippi and Loveluck 2016).

The source code designed by the platform developers dictates what kind of a consensus algorithm the network uses to maintain integrity and how the mining incentives are configured (Kroll et al. 2013; Gasser et al. 2015; Filippi and Loveluck 2016). As the profitability of the miners' business operations crucially depend on these factors, the platform developers have some output control over the miners (Torpey 2016a, b). The nodes also run the code designed by the platform developers, and therefore they are also subject to the output control, even if they are not so crucially dependent on the decisions made (Gasser et al. 2015).

When the platform developers write new versions of the network's client software, they can propose three kinds of alterations: ones that add new consensus rules to the network (*soft forks*), ones that remove or replace old consensus rules from the network (*hard forks*), and ones that do not affect the consensus protocols one way or the other. The more drastic the proposed changes to the consensus rules are, the more difficult it is to get them approved (Croman et al. 2016).

The users exercise output control over the miners in the sense that they provide the demand and the free market competition environment for the miners' services. A miner must be able to perform a competitive amount of computational work for a set market price in order to turn a profit.

Input control, or *gate-keeping*, is the enforcement of some pre-defined, objective criteria as a prerequisite for granting entry into the platform ecosystem. Typically the term has been used in reference to the platform owner exercising control over what kinds of application developers it allows into its cohort of complementary asset providers (Cardinal 2001; Evans et al. 2006; Boudreau 2010; Tiwana 2014). While no such control exists in the Bitcoin platform in this sense (Parker et al. 2016), input control is at play within the internal governance mechanics of the Bitcoin network itself.

The input control in Bitcoin is mainly exercised by the miners and the nodes, both of whom run the software developed by the platform developers (see Picture 2). As miners and nodes are free to decide which versions of the platform software they want to use, they are very influential as groups in determining what development features are accepted as a part of the platform and which ones are not (Kroll et al. 2013; Atzori 2015; Filippi and Loveluck 2016).

The miners also have input control toward users in the sense that they have the power to decide which pending transactions are entered into the blockchain and which ones are not (Dimitri 2017). This form of control also has some characteristics of output control. The miners are limited in their capacity to add transactions into the blockchain by the block size specified in the consensus rules of the network. As the pending transactions have tips called *transaction fees* attached to them by the users, the miners have an incentive to attach the highest bidding transactions to the blocks first (Dimitri 2017; Kroll et al. 2013; Catalini and Gans 2016). This creates a free market competition environment where the users must provide adequate compensation for the service of the miners in order to have their service request fulfilled over other service requests.

Another form of input control in Bitcoin is that exercised by the nodes over the miners. As the nodes are effectively in charge of enforcing the consensus protocol of the network, they as a group have the power to decide which blocks proposed by the miners are accepted as a new part of the blockchain and which ones are rejected.

Process control pertains to rewarding and/or penalizing application developers for following prescribed methods and procedures of development. This can be so as to ensure interoperability with the rest of the platform, for example (Tiwana 2014).

Process control is exercised in the Bitcoin network by users, application providers, and to some extent also miners (see Picture 3).

The platform developer community receives funding for their work from the application providers who have built their businesses on top of the Bitcoin platform and therefore depend on its development. This gives application providers some

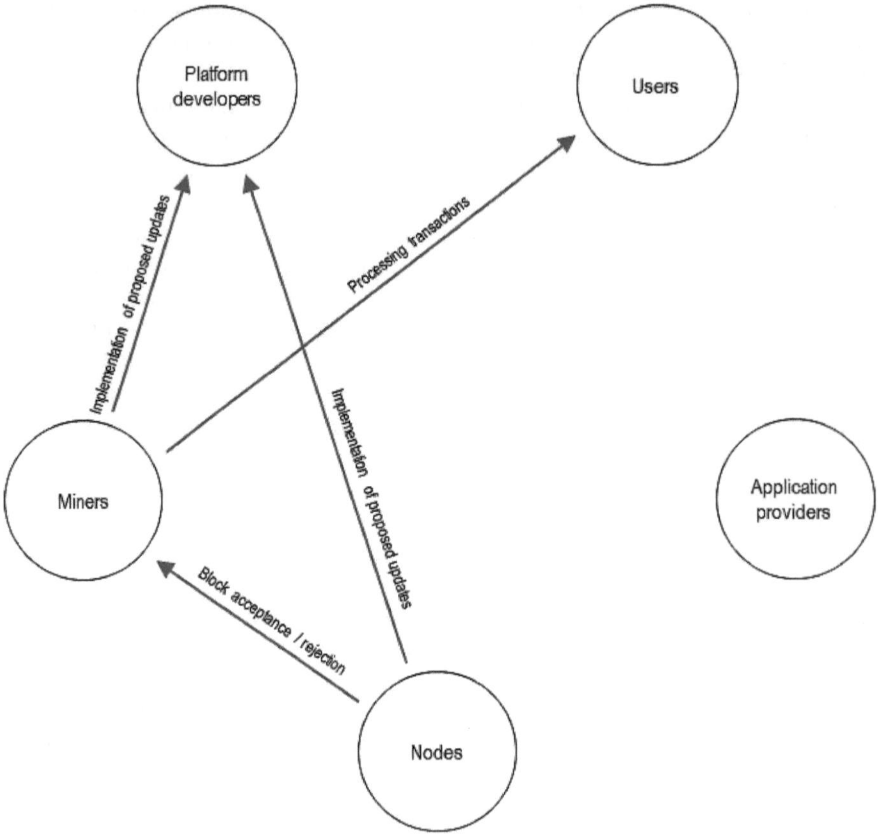

Picture 2 Input control in the Bitcoin platform

leverage over platform developers in determining what kinds of features should be incorporated into the source code and how the development should take place.

The same kind of a situation applies to nodes. Nodes are not compensated for their efforts through revenue split models in the same way as miners are, so they depend entirely on the good will of their respective owners to be set up and to have their operational costs covered. For this reason, the most reliable nodes of the network are usually maintained by parties who have the strongest vested interest to pay for them—that is, the market sides with the strongest ability to profit directly from participating in the platform: the application providers and miners (see Hagiu 2014). This gives them some degree of process control over what kinds of software updates the nodes will support, for example.

Perhaps the strongest position of process control, however, is held by the users over the miners. The motivation of the miners for participating in the platform is to make profit through transaction fees and block rewards, both paid out in the cryptocurrency tokens native to the platform. As stated above, the miners and the nodes

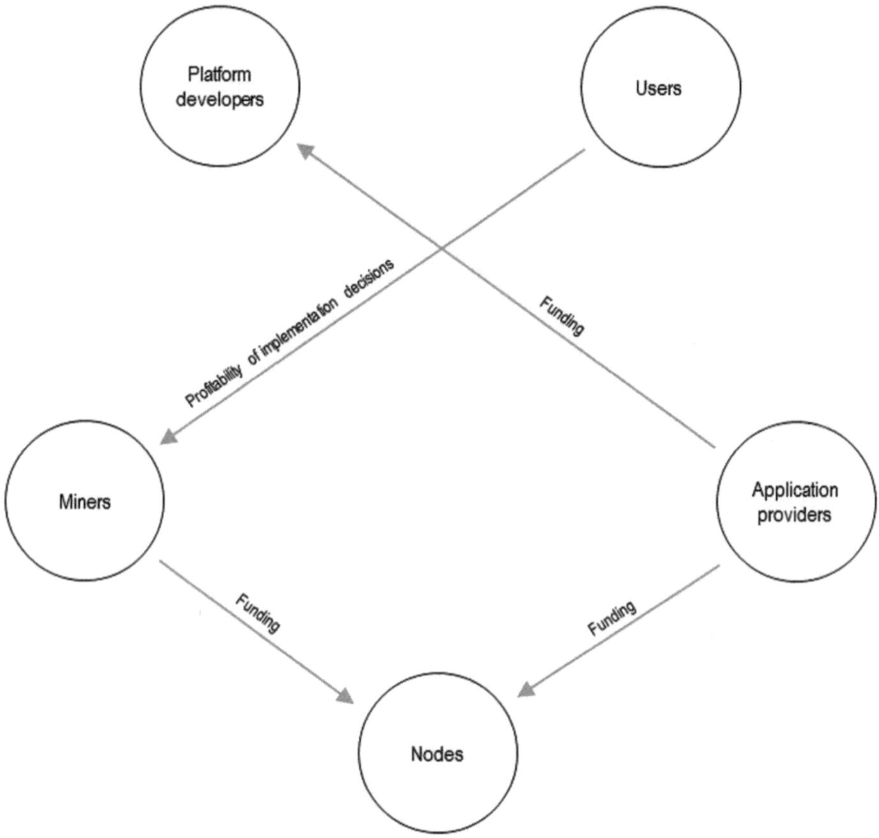

Picture 3 Process control in the Bitcoin platform

have input control over the platform developers in deciding which software updates get implemented. However, if the miners choose to implement software code alterations that are not to the liking of the users, they will sell off their cryptocurrency tokens, lowering the exchange rate of the cryptocurrency, which in turn will directly affect the mining profitability (Athey et al. 2016). The same process control mechanism also discourages minorities of the miners from implementing consensus-breaking software updates, or so-called hard forks, which would fragment the network into a larger number of smaller networks (Reijers et al. 2016).

Relational control or *clan control* manifests itself in the shared norms and values held by the platform participants (Ouchi 1979; Kirsch 1997; Tiwana 2014).

The cryptocurrency tokens used in blockchain-enabled platforms provide a strong mechanism for *relational control* (see Picture 4). They can be freely exchanged with conventional currencies and thus acquired by anyone who so pleases (Athey et al. 2016). The cryptocurrency serves as a medium of internal and external cooperation in the platform ecosystem, and it facilitates the incentivization

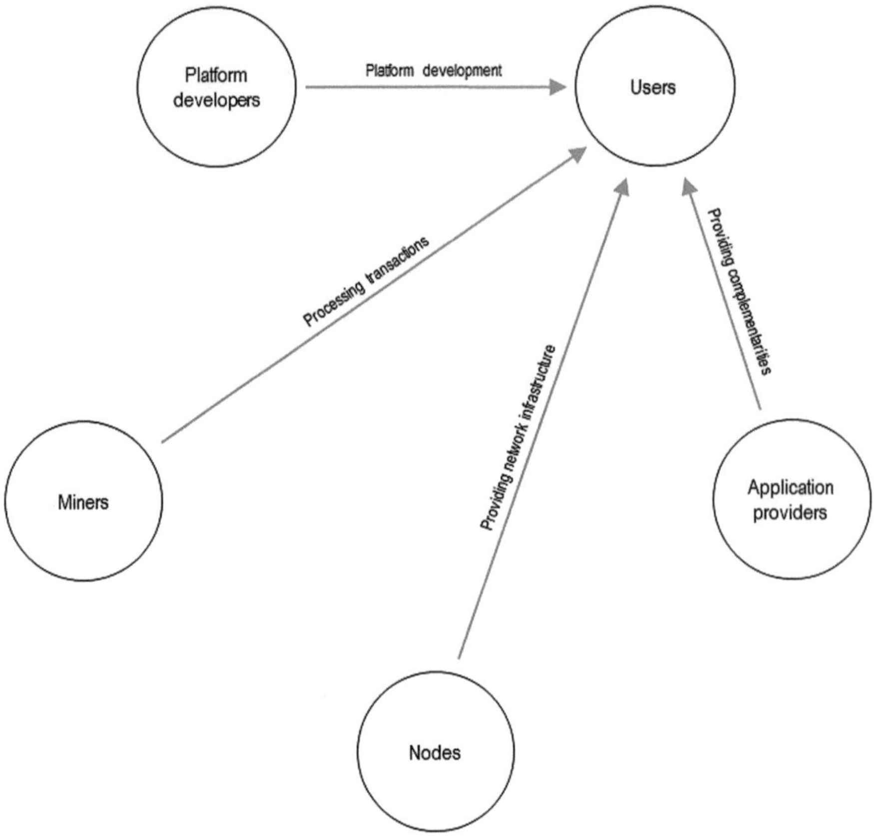

Picture 4 Relational control in the Bitcoin platform

of various participants by enabling a split revenue model for open participation (Catalini and Gans 2016; Böhme et al. 2015).

As the tokens circulated within the ecosystem are scarce, and as the transactions within the platform are settled in these tokens, increased activity in the platform ecosystem increases their demand, as described by the quantity theory of money:

$$M \times V = p \times q$$

where

M = the total amount of cryptocurrency tokens in circulation
V = the velocity of circulation
p = price level in the platform ecosystem
q = financial activity in the platform ecosystem

The rules and restrictions on minting new cryptocurrency tokens are governed by the source code of the client software on which the blockchain network operates.

Therefore, how and by whom these tokens can be minted varies from system to system. Usually, however, the minting of new tokens is algorithmically restricted, and therefore the total amount of tokens (*M*) is practically finite at any given moment (Böhme et al. 2015).[5]

The velocity (*V*) at which cryptocurrency tokens circulate in the ecosystem is limited by two factors: the average transaction value and the transaction throughput capacity of the network. The average transaction value is dictated by the types of financial activity within the platform ecosystem. Therefore, unless major changes occur in the ways and in the purpose for which the platform is utilized, the average transaction value can be expected to remain relatively stable.

Blockchain ledgers usually rely on a method where a new standard-sized block is added to a blockchain at more or less frequent intervals. While the throughput can be increased by a majority decision among the participants to increase the block size, under normal operating conditions, the capacity can be assumed to be fixed.

As financial activity on the platform (*q*) increases, it then follows from these assumptions that the only way for the equation to hold true is if the price level (*p*) decreases—that is, the value represented by each token must rise. Through this dependency and its projected growth development, the cryptocurrency tokens provide an investment vehicle which reflects the amount of activity on the platform ecosystem as directly monetizable value. Therefore, anyone in possession of these cryptocurrency tokens will find their goals and values aligned toward fostering the growth and the network effects of the platform ecosystem as much as possible (Alabi 2017; Catalini and Gans 2016; Athey et al. 2016).

Pricing Policies

Since blockchain-enabled distributed platforms do not have a centralized platform provider, they are not prone to monopoly behavior with their pricing policies, as is often the case with contemporary platforms (Parker et al. 2016).

The Bitcoin cryptocurrency platform uses revenue split schemes to incentivize collaboration among the platform providers. Partially these revenue splits are determined by free market competition within the market sides engaged in platform provision. Some revenue splits are determined algorithmically in the source code and the consensus protocols of the network.

As mentioned earlier, maintaining consensus in blockchain architectures requires miners to produce a proof-of-work. This can be characterized as a cryptographic testimony to the fact that the blockchain maintained by the network is authentic. As producing the cryptography required for the testimony requires computational

[5] For example, the first and most widespread blockchain platform to date, the Bitcoin network, uses native tokens called "bitcoins," each divisible to 100 million "satoshis." The source code of the system is set up in such a way that minting new bitcoins becomes exponentially more difficult as time passes on. This way, the maximum number of bitcoins that can exist in the system is limited to 21 million by the current consensus protocol.

work, miners typically receive compensations for sacrificing electricity for the good of the network. These compensations are commonly referred to as *mining rewards* or *block rewards* (Dimitri 2017; Böhme et al. 2015).

Mining rewards are the most prominent form of revenue splitting in blockchain architectures. The rewards are issued by minting new cryptocurrency into the system at frequent intervals. This increases the total supply of tokens which—according to the quantity theory of money—reduces the value of each token respectively (see section "Control"). Thus, the mining rewards somewhat resemble algorithmic *seigniorage*—an inflation tax collected from all platform participants and used to subsidize the production of public goods—namely, distributed consensus and immutability of record (Athey et al. 2016; Catalini and Gans 2016).

Miners also typically receive another form of compensation for their efforts. When a user wishes to make a transaction through the platform, they can add a voluntary *transaction fee* to their request. As the miners are limited in their capacity to add transactions to the blockchain due to the fixed block size and proof-of-work requirements, the transaction fees serve to ensure that the miners are incentivized to handle the transactions in an expedited manner. As most transactions have fees of some quantity attached to them by the users, the size of the transaction fee required for a normal throughput is determined by free market competition between the transaction requests (Dimitri 2017).

Unlike the miners, the platform developers and the nodes are not compensated for their platform provision functions through inflationary taxes and transaction tips (Filippi and Loveluck 2016). Instead, in order to get compensated, they must rely on a third form of revenue splitting quintessential to blockchain platforms: *token investments*, as described in section "Control".

Conclusions

In this paper, blockchain architectures were analyzed with the intent to determine whether they constitute multi-sided platforms. The paper also made an effort to delineate the internal governance structure of blockchain-enabled distributed platforms. This analysis was performed by applying Tiwana's (2014) framework on platform governance to the case examination of the Bitcoin cryptocurrency network.

On the basis of the case analysis of the Bitcoin cryptocurrency, blockchain architectures can clearly exhibit all the characteristics of a multi-sided platform, as outlined in the platform literature. The case architecture demonstrates direct network effects, as well as cross-side externalities. The interactions around the examined system can be described as multi-sided switch-role markets, and the technical and social boundary resources of these architectures are clearly defined.

Some discrepancies were observed, however, in regards to the wider perceptions of platforms in the platform literature. While blockchain-based distributed platforms coordinate and regulate the connections between its ecosystem participants

(Gawer and Cusumano 2002), they do not necessarily function as licensing authorities, as characterized by Rochet and Tirole (2003). To a certain degree, distributed platforms serve the role of a public interest regulator but do not exhibit the characteristics of a monopolist platform owner, as described by Boudreau and Hagiu (2008).

Several conclusions can be drawn from the case analysis in section "Mechanisms of Internal Governance". Firstly, in blockchain-enabled distributed platforms, not only have the product and service innovations pertaining to the platform been externalized, but the entire platform provision has been distributed across various market sides. In the examined case example of Bitcoin cryptocurrency, the distributed collaboration is held together by an intricate web of monetary incentives and different forms of interlocking control mechanisms exerted by the participating market sides toward one another.

Secondly, although Bitcoin is technically anarchistic in governance, some de facto structure exists to how the platform decision rights are allocated in Bitcoin. As no formal decision-making protocols are in place, an informal negotiation process takes place through various forms of indirect signaling.

Thirdly, blockchain-enabled distributed platforms seem to have introduced a new method of platform monetization. The contemporary platform business models have mainly been based on enabling direct interactions between the different market sides and monetizing by controlling access to those interactions and by leveraging information asymmetries. In blockchain-enabled distributed platforms, however, the linchpin business model seems to revolve around launching an independent, self-sustained open ecosystem of direct interactions and monetizing on the tokens of value utilized as the means of exchange in that ecosystem.

Fourthly, much like contemporary multi-sided platforms, blockchain-enabled distributed platforms also seem to employ split revenue schemes, with two notable differences in regard to conventional multi-sided platforms. The first difference is that in Bitcoin, the revenue splits are not used to share profits between external complementary asset providers and the platform's owner but between all the different market sides participating in the platform provision and the ecosystem at large. The second difference is that the platform pricing and the split revenue schemes are not monopolized by any single party but rather determined by the quasi-anarchistic, semi-structured decision-making process between the market sides, as well as free market competition mechanics.

Boudreau and Hagiu (2008) have hypothesized that all the coordination problems related to the collaboration around multi-sided platforms cannot be solved through mere price-setting alone. In contemporary platforms, the economic incentivization only prevents market failures to the extent that the platform participants are compensated for their participation through revenue split models. In blockchain-enabled distributed platforms, however, the economic incentivization is more equilateral and ubiquitous than in contemporary platform models. Therefore, we argue that the variety of coordination problems and market failures that *can* be addressed through price setting is quite likely to be much wider in scope in distributed platforms (see Catalini and Gans 2016).

Acknowledgments This paper has been drafted as a part of the Work and Wealth in the Era of Digital Platforms (BRIE-ETLA) research collaboration of 2015–2018. Timo Seppälä has also received financial support for his work from the Blockchains Boosting Finnish Industry (BOND) research project.

References

Alabi, K. (2017, June). Digital blockchain networks appear to be following Metcalfe's Law. *Electronic Commerce Research and Applications*. Available at https://doi.org/10.1016/j.elerap.2017.06.003

Armstrong, M. (2006). Competition in two-sided markets. *RAND Journal of Economics, 37*(3), 668–691. Available at http://onlinelibrary.wiley.com/doi/10.1111/j.1756-2171.2006.tb00037.x/abstract

Aspers, P. (2008). Analyzing order: Social structure and value in the economic sphere. *International Review of Sociology, 18*(2), 37–41.

Athey, S., Parashkevov, I., Sarukkai, V., & Xia, J.. (2016, August 1). *Bitcoin pricing, adoption, and usage: Theory and evidence*. Stanford University Graduate School of Business Research Paper No. 16–42. Available at SSRN: https://ssrn.com/abstract=2826674

Atzori, M. (2015). *Blockchain technology and decentralized governance: Is the state still necessary?* Available at *SSRN*, pp. 1–37. Available at papers.ssrn.com

Baldwin, C.Y. & Woodard, C.J., 2009. The architecture of platforms: A unified view. In A. Gawer, ed. Platforms, markets and innovation. Cheltenham: Edward Elgar Publishing, 19–44.

Ballon, P. (2009a). *Platform types and gatekeeper roles: The case of the mobile communications industry*. Paper presented at the Druid Summer Conference.

Ballon, P. (2009b). The platformisation of the European mobile industry. *Communications and Strategies, 75*, 15.

Ballon, P., & Van Heesvelde, E. (2011). ICT platforms and regulatory concerns in Europe. *Telecommunications Policy, 35*(8), 702–714.

Baran, P. (1964). On distributed communications networks. *IEEE Transactions on Communications Systems, 12*(1), 1–9.

Bester, H., & Krähmer, D., (2008). Delegation and incentives. *RAND Journal of Economics, 39*(3), 664–682. Available at http://onlinelibrary.wiley.com/doi/10.1111/j.1756-2171.2008.00033.x/abstract

Böhme, R. et al. (2015). Bitcoin: economics, technology, and governance. *Journal of Economic Perspectives, 29*(2), 213–238. Available at https://doi.org/10.1257/jep.29.2.213%0Ahttp://pubs.aeaweb.org/doi/10.1257/jep.29.2.213

Boudreau, K. J., & Hagiu, A. (2009). Platform rules: Multi-sided platforms as regulators. In A. Gawer (Ed.), *Platforms, markets and innovation*. Cheltenham: Edward Elgar Publishing.

Boudreau, K. (2010). Open platform strategies and innovation: Granting access vs. devolving control. *Management Science, 56*(10), 1849–1872. Available at https://doi.org/10.1287/mnsc.1100.1215

Boudreau, K.J. & Hagiu, A., 2008. Platform rules: Multi-sided platforms as regulators. In A. Gawer, ed. Platforms, markets and innovation. Cheltenham: Edward Elgar Publishing, 1–29. Available at https://doi.org/10.2139/ssrn.1269966

Caillaud, B., & Jullien, B. (2003). Chicken & egg: Competition among intermediation service providers. *The RAND Journal of Economics, 34*(2), 309–328. Available at http://www.jstor.org/stable/1593720

Cardinal, L. B. (2001). Technological innovation in the pharmaceutical industry: The use of organizational control in managing research and development. *Organization Science, 12*(1), 19–36. Available at http://pubsonline.informs.org/doi/abs/10.1287/orsc.12.1.19.10119

Catalini, C., & Gans, J. (2016). *Some simple economics of the blockchain some simple economics of the blockchain*. Available at: https://papers.ssrn.com/sol3/papers.cfm?abstract_id=2874598
Cawrey, D. (2014). What are bitcoin nodes and why do we need them? *CoinDesk*. Available at https://www.coindesk.com/bitcoin-nodes-need/
Consensus. (2017). *Can bitcoin scale? Panel discussion*. New York: CoinDesk.
Croman, K. et al. (2016). On scaling decentralized blockchains. In *Proceeding of the 3rd Workshop on Bitcoin and Blockchain Research*. Available at http://fc16.ifca.ai/bitcoin/papers/CDE+16
Cusumano, M. A., & Selby, R. W. (1995). *Microsoft secrets: How the world's most powerful software company creates technology, shapes markets and manages people*. New York: Free Press.
Cusumano, M. A., & Yoffie, D. (1998). *Competing on internet time: Lessons from netscape and its battle with Microsoft*. New York: Free Press.
Dahlander, L., & Wallin, M. (2006). A man on the inside: Unlocking communities as complementary assets. *Research Policy, 35*, 1243–1259.
De Filippi, P., & Loveluck, B. (2016). The invisible politics of bitcoin: Governance crisis of a decentralised infrastructure. *Internet Policy Review*, *5*(3), 1–28. Available at http://policyreview.info/articles/analysis/invisible-politics-bitcoin-governance-crisis-decentralised-infrastructure
de Reuver, M., Sørensen, C., & Basole, R. C. (2017). The digital platform: A research agenda. *Journal of Information Technology*. https://link.springer.com/journal/41265/onlineFirst/page/1.
Dimitri, N. (2017). Bitcoin mining as a contest. *Ledger Journal, 2*(1), 31–37.
Eisenhardt, K. M. (1989). Building theories from case study research. *The Academy of Management Review, 14*(4), 532–550.
Eisemann, T. (2008). Managing shared and proprietary platforms. *California Management Review, 50*(4), 31–53.
Eisenmann, T., Parker, G., & van Alstyne, M. (2006). Strategies for two-sided markets. *Harvard Business Review, 84*(10), 1–10.
Evans, D. S., Hagiu, A., & Schmalensee, R. (2006). *Invisible engines: How software platforms drive innovation and transform industries*. Cambridge, MA: MIT Press. Available at https://mitpress.mit.edu/sites/default/files/titles/content/9780262050852_Download_the_full_text.pdf
Gasser, U., Budish, R., & West, S. M. (2015). *Multistakeholder as Governance Groups*,
Gawer, A. (2009). *Platforms, markets and innovation*. Cheltenham: Edward Elgar.
Gawer, A., & Cusumano, M. A. (2002). *Platform leadership* (p. 316). Boston: Harvard Business School Press.
Gawer, A., & Henderson, R. (2007). Platform owner entry and innovation in complementary markets: Evidence from intel. *Journal of Economics and Management Strategy, 16*(1), 1–34.
Ghazawneh & Henfridsson. (2013). Balancing platform control and external contribution in third-party development: The boundary resources model. *Information Systems Journal, 23*, 173–192.
Gilson, L. L., & Goldberg, C. B. (2015). Editors' comment: So, what is a conceptual paper? *Group & Organization Management, 40*(2), 127–130.
Hagiu, A. (2014). Strategic decisions for multisided platforms. *MIT Sloan Management Review, 55*(2), 71–80. Available at http://sloanreview.mit.edu/article/strategic-decisions-for-multisided-platforms/
Hagiu, A., & Wright, J. (2015). Multi-sided platforms. *International Journal of Industrial Organization, 43*, 162–174. Available at https://doi.org/10.1016/j.ijindorg.2015.03.003
Katz, M. L., & Shapiro, C. (1994). Systems competition and network effects. *Journal of Economic Perspectives, 8*(2), 93–115. Available at https://www.aeaweb.org/articles?id=10.1257/jep.8.2.93
Kim, D., & Kogut, B. (1996). Technological platforms and diversification. *Organization Science, 7*(3), 283–301.
Kirsch, L. J. (1997). Portfolios of control modes and IS project management. *Information Systems Research, 8*(3), 215–239. Available at http://pubsonline.informs.org/doi/abs/10.1287/isre.8.3.215
Kroll, J. A., Davey, I. C., & Felten, E. W. (2013). The economics of bitcoin mining or bitcoin in the presence of adversaries. In *Proceedings of WEIS* (pp. 1–21).

Laszka, A., Johnson, B., & Grossklags, J. (2015). When bitcoin mining pools run dry. A game-theoretic analysis of the long-term impact of attacks between mining pools. In *International Conference on Financial Cryptography and Data Security* (pp. 63–77). Berlin.

Nakamoto, S. (2008). *Bitcoin: A peer-to-peer electronic cash system.* www.bitcoin.org, p. 9. Available at https://bitcoin.org/bitcoin.pdf

Ouchi, W. G. (1979). A conceptual framework for the design of organizational control mechanisms. *Management Science, 25*(9), 833–848. Available at https://business.illinois.edu/josephm/ba545_fall2011/s3/ouchi%20(1979).pdf

Parker, J., & van Alstyne, M. (2005). Two-sided network effects: A theory of information product design. *Management Science, 51*(10), 1494–1504.

Parker, G., & van Alstyne, M. W. (2014). *Innovation, openness & platform control.* Available at https://ssrn.com/abstract=1079712

Parker, G. G., van Alstyne, M. W., & Choudary, S. P. (2016). *Platform revolution: How networked markets are transforming the economy and how to make them work for you.* New York: WW Norton & Company.

Pon, B., Seppälä, T., & Kenney, M. (2014). Android and the demise of operating system-based power: Firm strategy and platform control in the post-PC world. *Telecommunications Policy, 38*(11), 979–991.

Reijers, W., Brolcháin, F. O., & Haynes, P. (2016). Governance in blockchain technologies & social contract theories. *Ledger Journal, 1*(1), 134–151.

Rochet, J.-C., & Tirole, J. (2003). Two-sided markets: A progress report. *The RAND Journal of Economics, 37*(3), 645–667.

Rochet, J.-C., & Tirole, J. (2006). Two-sided markets: A progress report. *The RAND Journal of Economics, 37*(3), 645–667. Available at: http://www.jstor.org.ez.urosario.edu.co/stable/25046265.

Rysman, M. (2009). The economics of two-sided markets. *Journal of Economic Perspectives, 23*(3), 125–143. Available at https://www.aeaweb.org/articles?id=10.1257/jep.23.3.125

Schilling, M. A. (2005). *Strategic management of technological innovation.* Boston: Tata McGraw-Hill Education.

Srinivasan, B. S., & Leland, L. (2017). *Quantifying decentralization.* Available at: https://news.21.co/quantifying-decentralization-e39db233c28e

Sutton, R. I., & Staw, B. M. (1995). What theory is not. *Administrative Science Quarterly, 40*(3), 371–384.

Teece, D. J. (1986). Profiting from technological innovation: Implications for integration, collaboration, licensing and public policy. *Research Policy, 15*(6), 285–305. Available at http://www.sciencedirect.com/science/article/pii/0048733386900272

Teece, D. J. (1988). Capturing value from technological innovation: Integration, strategic partnering, and licensing. *Interfaces, 18*(3), 46–61.

Tiwana, A. (2014). *Platform ecosystems: Aligning architecture, governance, and strategy.* Waltham: Morgan Kaufmann.

Tiwana, A., Konsynski, B., & Bush, A. (2010). Platform evolution: Coevolution of platform. *Information Systems Research, 21*(4), 675–687. Available at http://pubsonline.informs.org/doi/abs/10.1287/isre.1100.0323

Torpey, K. (2016a). The checks and balances of bitcoin governance. *Bitcoin Magazine.* Available at https://bitcoinmagazine.com/articles/the-checks-and-balances-of-bitcoin-governance-1454695089/

Torpey, K. (2016b). Why miners are lowest in the hierarchy of control over bitcoin. *Coin Journal.* Available at http://coinjournal.net/why-miners-are-lowest-in-the-hierarchy-of-control-over-bitcoin/

van Alstyne, M., Parker, G., & Choudary, S. (2016). Pipelines, platforms, and the new rules of strategy. *Harvard Business Review*, April 2016, pp. 54–63. Platform revolution: How networked markets are transforming the economy--And how to make them work for you.

van Wirdum, A. (2016). Who funds bitcoin core development? How the industry supports bitcoin's reference client. *Bitcoin Magazine*. Available at https://bitcoinmagazine.com/articles/who-funds-bitcoin-core-development-how-the-industry-supports-bitcoin-s-reference-client-1459967859/

Vasek, M., Thornton, M., & Moore, T. (2014). Empirical analysis of Denial-of-Service Attacks in the Bitcoin ecosystem. In R. Böhme et al. (Eds.), *International conference on financial cryptography and data security* (pp. 57–71). Barbados: Christ Church.

Yli-Huumo, J. et al., 2016. Where is current research on blockchain technology?—A systematic review. *Plos One, 11*(10), e0163477. Available at http://dx.plos.org/10.1371/journal.pone.0163477

Yoffie, D., & Kwak, M. (2006). With friends like these: The art of managing complementors. *Harvard Business Review*, (September), 88–98.

Yoo, Y., Henfridsson, O., & Lyytinen, K. (2010). The new organizing logic of digital innovation: an agenda for information systems research. *Information Systems Research, 21*, 724–735.

Open Access This chapter is licensed under the terms of the Creative Commons Attribution 4.0 International License (http://creativecommons.org/licenses/by/4.0/), which permits use, sharing, adaptation, distribution and reproduction in any medium or format, as long as you give appropriate credit to the original author(s) and the source, provide a link to the Creative Commons license and indicate if changes were made.

The images or other third party material in this chapter are included in the chapter's Creative Commons license, unless indicated otherwise in a credit line to the material. If material is not included in the chapter's Creative Commons license and your intended use is not permitted by statutory regulation or exceeds the permitted use, you will need to obtain permission directly from the copyright holder.

Value Creation from the Internet of Things in Heavy Machinery: A Middle Manager Perspective

Marko Sommarberg, Robin Gustafsson, Zeerim Cheung, and Eero Aalto

Abstract We develop a novel understanding of value creation from the Internet of Things (IoT) in the heavy machinery industry. We analyze the operational middle management perspective on the current state of IoT implementation in four industry segments utilizing heavy machinery, including chemicals, electrical equipment, marine, and pulp and paper. We find four value drivers from IoT (product optimization, maintenance and recovery optimization, energy efficiency, and safety improvements) that pertain to the underlying joint value driver of predictability with operating heavy machinery. Furthermore, we outline current core issues and constraints in IoT data utilization and value realization in the heavy machinery industry. Our findings expose how middle managers recognize high value potential from IoT implementation but indicate only gradual value capture if data gathering, knowledge sharing, and data analytics are not improved. Architectural transformation programs that include establishing knowledge-sharing hubs and connected expertise between organizational units, external experts, and suppliers are needed to unlock the full novel value-creation potential of the IoT in heavy machinery.

Keywords Value creation · Internet of things · Heavy machinery · Middle manager

M. Sommarberg
ABB Industrial Automation Division, Helsinki, Finland

R. Gustafsson (✉) · Z. Cheung · E. Aalto
Department of Industrial Engineering and Management, Aalto University School of Science, Espoo, Finland
e-mail: robin.gustafsson@aalto.fi

© Springer Nature Singapore Pte Ltd. 2018

A. Smedlund et al. (eds.), *Collaborative Value Co-creation in the Platform Economy*, Translational Systems Sciences 11,
https://doi.org/10.1007/978-981-10-8956-5_11

Introduction

The Internet of Things (IoT) is emerging as a novel source of value creation in heavy machinery. IoT technologies allow for the real-time connectivity and interconnection of devices in the cloud, which generates aggregated data from the application and usage of the devices through sensors (e.g., Porter and Heppelmann 2014; Atzori et al. 2017). IoT technologies are placed in the machinery and production facilities. IoT holds the potential to change the underlying economics of processes and operations of those firms that utilize and deliver heavy machines, technologies, and services. The industry architecture is defined as the division of labor between firms in an industry encompassing the physical and technical architecture (Jacobides et al. 2006). The architectures of these industries are currently in a state of flux where new business models, new types of relationships between firms, and novel entrant engagements are being explored. In recent years, the enabling technological conditions for IoTs (including sensors, communication, software, and data analytics) have become more favorable with decreasing costs and performance advancements (Atzori et al. 2010, 2017; Evans and Annunziata 2012).

The IoT allows industrial firms to enhance their existing value creation through increased productivity. The IoT allows for the capturing of data from machines, aggregating information across networks and either taking immediate action based on the data or gradually learning more about the processes (Manyika et al. 2015). This allows firms to remotely control and monitor machines and factory operations, increase automatization, and optimize machine use, factory operations, inventory, and supply chain management. Novel, big datasets, real-time connectivity, data analytics, and processing algorithms have the potential to create intelligent production and machine systems. Machine monitoring, control, and optimization can become increasingly autonomous, where machines learn about their environment, self-diagnose service needs, and adapt to a user's preferences (Porter and Heppelmann 2014). Overall, the efficiency gains through which the IoT can realize are increased production uptime, improved asset utilization, decreased energy consumption, better operational safety, and transparent, coherent processes.

The IoT in heavy machinery applications changes existing B2B industrial customer relationships from turn-key solution deliveries and after-sales service relationships toward an "online relationship." The suppliers and customers are becoming more closely connected to each other through data and remote connections (Porter and Heppelmann 2014). In-depth knowledge of customer systems enables suppliers to service their customers in real time, offer additional products and services, and customize offerings. In addition, the closer supplier-customer relationship has the potential to increase cooperation in product development and create faster product development cycles. Understanding value creation processes from the customer point of view allows firms to generate greater levels of customer satisfaction, loyalty, and repurchase behavior and to create a lock-in to the supplier (Bradley et al. 2013).

Middle managers are defined as "any manager two levels below the CEO and one level above line workers and professionals" (Huy 2001: 73). They are critical in IoT adaptation and implementation. They have a strong understanding and knowledge of the firm operations, access to the top management, and unique positions to leverage informal networks of the firm (Huy 2001; Balogun 2003). Therefore, middle managers can bridge continuity and change (Huy 2001). In addition, they are central actors that interpret and sell change in the firm (Rouleau 2005). They are intermediates between disconnected actors, domains, levels, and units in the firm and act as mediators between firm strategy and day-to-day activities (Wooldridge et al. 2008). The intermediating role includes influencing outcomes and interpreting strategy both upward and downward in an organizational hierarchy, thus making middle managers important actors in both strategy formulation and implementation.

IoT solutions have not yet been widely adopted and capitalized upon (Accenture 2015b), even though the substantial long-term benefits are acknowledged by top management (Koch et al. 2014; PwC 2015), and the enabling technological conditions already exist (Atzori et al. 2017). Currently, only a fraction (approximately 1%) of existing IoT data is utilized (Manyika et al. 2015). The early technological applications of the IoT have been put in place in many industries, and top management vision is supporting further advancements. Further utilization of the potential of the IoT has more to do with operational and organizational practice transformation and strategy implementation than the technological conditions or generic strategic vision shared by top managers. As such, there is limited understanding on how the data gathered from the machinery could enhance value creation from the perspective of middle managers. Furthermore, there are inadequate foundations for implementing the IoT-related systems, practices, and services. Hence, in order to study IoT-based value creation in the heavy machinery industry, we assess the IoT-related key value realization constraints and drivers from the perspective of middle managers. This allows us to both uncover the current middle management perspective of IoT and formulate managerial implications for future IoT strategies in the heavy machinery industry.

Study, Data, and Methods

We study the middle management perspective on value creation from IoT in the context of four industry segments utilizing heavy machinery (chemicals, electrical equipment, marine, and pulp and paper). Our data sample includes one sample firm from each of the studied industry segments (in total four firms) in which heavy machinery is critical to value creation. The case firms were deliberately chosen because the firms had not implemented full-scale IoT solutions, but they had interest in and plans for a wider implementation of and more investments in the IoT. Since there is a limited amount of research on IoT implementation and especially on the middle manager perspective on IoT value creation, we chose an inductive qualitative study design aimed at building a new theoretical understanding rather than

Table 1 List of interviewees

Industry	Area of responsibility
Chemicals	Maintenance manager
	Electrical maintenance manager
	IT manager
	Electricity and automation engineer
	Electricity work planner
Electrical equipment	Unit manager, unit A
	Unit manager, unit B
	Plant manager, production manager
	Operations manager
	Quality manager
Marine	Captain
	Chief engineer
	Operations manager
	Electricity manager
Pulp and paper	Plant manager
	Production manager
	Maintenance manager
	IT manager, factory A
	IT manager, factory B
	Development manager
	Planning manager

testing or elaborating on constructs (Gioia et al. 2013). Our approach seeks to obtain a novel understanding of the IoT value drivers in heavy machinery by engaging with the organizational actors experiencing the phenomenon of interest. Using semi-structured interviews, in 2016, we engaged with middle management (operations and product management) in each of the case firms. We conducted a total of 21 interviews (for a list of interviewees, their positions, and the respective case firms' industries, see Table 1).

The interviews were conducted in Finland, although the middle managers' companies all operated in multiple countries. We complemented our interview data with observations by attending six IoT industry events during our data collection process, all during 2016.

We analyzed the data collected from the interviews with comprehensive qualitative coding using Atlas.TI. In the analysis process, we used the interview data to form first-order concepts. After careful deliberation with the research team, we grouped these into second-order themes that highlighted key dimensions arising from the interview data (Gioia et al. 2013). To verify our findings and build managerial implications, we continuously reanalyzed our results, the prior research, and the industry event observation data until additional rounds did not provide further insights. From this process, we were able to extract our main empirical findings in aggregated dimensions, including the present state of data use in value creation in

Table 2 Present state of data use in heavy machinery

Second-order themes	First-order concepts	Exemplary quotes related to the theme
Data abundance	High amount of data from automation	"We have basic measurements that we need in our automation system about flows, temperature, pressure, frequency and paper and board quality-related ones that tell, for example, square weights, humidity, calibers and fiber electrical attributes. We try to measure everything that is reasonably measurable." – *IT Manager, Factory A, Pulp and Paper*
	Too much information	"We already have too much information from the automation system to our operational employees, but, on the other hand, this information could be used better when disturbances occur so that diagnostics would tell the employees what has happened." – *IT Manager, Chemicals*
Existing data from machine operation underutilized	Current data could be used better in operating machines	"We know that we have a lot of data. We collect high amounts of data about the disturbances, about the reasons behind the disturbances. So, basically, we have all the data. However, when we have the database, how are we able to find something from there? The right things from the mass of data are the most important." – *Plant Manager, Pulp and Paper*
	Existing remote connections	"To some extent, we have remote connections in place. Sometimes, a supplier takes a remote connection, but it is not constant. If problems occur, we ask certain firms to take a look at them." – *Electricity and Automation Engineer, Chemicals*
	Data could be stored for a longer time for disturbance diagnostics	
	Lacking ways to intelligently explore data	
	Some applications of automated maintenance are available	
Fragmented data collection practices	Manual data collection	"Currently, data analysis is purely manual." – *Production Manager, Electrical Equipment*
	Employer data collection practices vary	"Our data analysis varies between employees and is done when a problem occurs, so it is not very comprehensive. It is more about finding the root causes of problems and maybe if we anticipate that problems might occur somewhere." – *Maintenance Manager, Chemicals*
	Data integration from sensors and machines lacking	

(continued)

Table 2 (continued)

Second-order themes	First-order concepts	Exemplary quotes related to the theme
Non-predictive maintenance	Time-based maintenance	"If something has happened, like a machine has shut down, then we go through the automation system data to try to figure out what the problem is." – *Operations Manager, Marine*
	Data analyzed after something has occurred	"We use data very poorly. We have the weakness that data are only researched when a failure occurs. Only then do we research the causes but not beforehand. We should be doing this more, and we should have resources that somebody would research this more deeply." – *Electrical Maintenance Manager, Chemicals*
	Maintenance program poorly followed	

heavy machinery (see Table 2), the value realization constraints in heavy machinery companies (see Table 3), and predictability as the key IoT value driver (see Table 4). We also stated some managerial implications for firms utilizing or planning to utilize the IoT in heavy machineries.

Findings

First, we present our findings regarding the current situation with regard to data use in firms using heavy machinery. The second part presents our findings regarding data and IoT value realization constraints in heavy machinery. The third and final part presents our findings of key IoT value drivers from the viewpoint of middle managers.

Present State of Data Use in Value Creation in Heavy Machinery

Our study exposes four core issues regarding the *state of data use in heavy machinery*: (1) data abundance, (2) existing data from machine operations underutilized, (3) fragmented data collection practices, and (4) reactive and preventive maintenance. Our findings are summarized in Table 2.

The middle managers we interviewed throughout our study expressed that there is already an abundance of data existing at factories and units using heavy machinery. All of the case firms we studied used automation systems in their production processes. The interviewees expressed that they measure almost everything from the process, including flows, temperatures, pressures, rotations, and vibrations. In contrast to the state-of-the-art automation systems, the case firms mostly manually

Table 3 Value realization constraints in heavy machinery companies

Second-order themes	First-order concepts	Exemplary quotes related to the theme
Organizational constraints	Organizational silos in factories and units	"Organization silos are one obstacle. We have quite limited areas of responsibility, and we do not look at the big picture much. The IoT and looking to the big picture is a big change compared to current operations. People have been looking at only their production line and not neighboring lines." – *Quality Manager, Electrical Equipment*
	Costs and paybacks for investments in the IoT hard to estimate (rationale unclear)	"We do not have time to do data analysis. It is engineers' work, and I am ashamed that there is no regular search for variances to see when things are changing in production. It would not be anything groundbreaking, but it is simply something that we would need to do but are still not doing. It would be great if it were automatized… [automatized data analysis] should be in place so that we would not need to go through trends manually, and I see that this would yield high benefits if the analysis was automatized." – *Production Manager, Pulp and Paper*
	Difficult to find time between production and maintenance	
	Managers and workers lack excitement about developing data utilization and the IoT	
Resource and capability constraints	Increased automation and reduction of heavy machine workers	"In troubleshooting, the more complicated the system is, the more it brings value to have an external expert that can connect remotely. In a sense, we have a limited ability to handle more complicated problems compared to the people who work day-to-day with similar matters and can fix the problem very quickly as long as they get the data. This has proven to be valuable." – *Chief Engineer, Marine*
	R&D and IT resource constraints	"Our maintenance organization is getting slimmer and slimmer and our production organization takes care of production more autonomously, but they do not think that much about the equipment condition….The slimmer our organizations are getting, the more we need external help." – *Electrical Maintenance Manager, Chemicals*
	Lacking data analysis capabilities lacking	"If we think about the whole IT function, all the experimentations have decreased, and so we do not test new ideas as much as before…Our equipment suppliers might do these activities more often, but in our organization, this is very little… We have been cutting costs for many years everywhere, so we cannot take risks and invest in something totally new to see if it works. This has clearly decreased." – *IT Manager, Factory B, Pulp and Paper*
	IT security challenges	"Our main suppliers [suppliers 1 and 2] could send data packages to us about our equipment condition, and they could do data analysis and send suggestions based on their findings." – *Captain, Marine*
	Constraints to engage external experts for analyzing machines (i.e., troubleshooting)	

(continued)

Table 3 (continued)

Second-order themes	First-order concepts	Exemplary quotes related to the theme
Sensemaking constraints regarding data use and the IoT	Managers feel that more data are not needed	"Current data could be used more. It would be useful that we would know how to use it better. It is very clear that it would help in developing different process states, speed and quality issues. Usually, the problem is not that we would not, have data but instead, how we could use the current data smarter." – *Unit Manager, Unit A, Electrical Equipment*
	Abstract investments are difficult to plan	"It is quite difficult to compare different factories because they are quite different. If the factories had similar production processes and products, then there might be some benefits. However, I see more internal benefits." – *Production Manager, Electrical Equipment*
	Difficult to compare factories Safety benefits from the IoT are unclear	

Table 4 Predictability as the key value driver from the IoT

Second-order themes	First-order concepts	Exemplary quotes related to the theme
Production optimization	Production quality improvements	"Where we would have use of big data and automatic data collection would be especially when we see that the production line is making the best quality end products, and then we could save those data and parameters easily. The next time in production, we could control the production so that it would alert if operators would try to break the optimal quality parameters. For this kind of monitoring and control, we would pay anything." – *Production Manager, Electrical Equipment*
	Cost-efficiency improvements	"The future should be in advanced process control where data are analyzed, and optimization programs are developed based on that, so that any process can be controlled." – *Maintenance Manager, Chemicals*
	Production reliability improvements	
	Available production improvements	
Maintenance and recovery optimization	Capabilities for predictive maintenance	"We should develop predictability about when our production equipment is about to break down. It would be interesting to know beforehand…If we could prolong all equipment maintenance to stoppages without surprises it could be very valuable…. Predicting that equipment lasts until the next stoppage would be useful so that we would not need to fix it too early. If that information would be available two to three weeks beforehand, it would be good, as then we would not need to shut down production for separate stoppages." – *Maintenance Manager, Pulp and Paper*
	Speeding up production recovery	"Machines can be sometimes extremely difficult to fix and if something comes up quickly. It is very beneficial to know the machine and the symptoms beforehand. With this kind of predictability, it is possible to get acquainted with the machine's history, repair history and know how it has been functioning, what it includes and be able to procure needed spare parts beforehand." – *IT Manager, Factory A, Pulp and Paper*
	Machine life cycle management	
Energy efficiency	Energy efficiency gains are high within and across factories and units (especially in energy-intensive sectors)	"There is a lot of potential in energy efficiency. In that area, we can benefit a lot from IoT systems, and then we could connect to different factories. This works definitely." – *Plant Manager, Pulp and Paper*
	Energy usage optimization	"We measure energy in many places already, but we do not use that information to control our production. I think it should control production and tell how much was consumed, making certain variety or comparing energy usage between shifts. There would be potential." – *Electrical Maintenance Manager, Chemicals*

(continued)

Table 4 (continued)

Second-order themes	First-order concepts	Exemplary quotes related to the theme
Safety improvements	Factory safety improvements	"In many cases, a reliable factory is also safe and cost-efficient when there is no constant need for repairs. Also, abnormal situations are the least safe for maintenance. In that sense, predictability and knowing equipment failures beforehand result in smaller disturbances and repairs in maintenance." – *Maintenance Manager, Chemicals*
	Predictability improves safety	"It could be summarized that fewer unexpected problems that need uncertain and quick fixes result in more predictability, so that the problems can be fixed in greater control, and this all leads to higher safety." – *Production Manager, Electrical Equipment*
	Advancing public image (safety and the IoT)	

conduct their data analyses. Therefore, the factories, units, and firms using heavy machinery have an abundance of data, but they are neither analyzed nor extensively used. As one production manager explained:

> Currently, data analysis is purely manual. – *Production Manager, Electrical Equipment*

Despite the abundance of data, the factories, units, and the corporation are not able to use those data in the ways that they would like. Existing data from machine operations are clearly underutilized. The corporations lack ways to intelligently explore the data. As one plant manager explained to us:

> We know that we have a lot of data. We collect high amounts of data about the disturbances and about the reasons behind the disturbances. So basically, we have all the data. However, when we have the database, how are we able to find something from there? The right things from the mass of data are the most important. – *Plant Manager, Pulp and Paper*

One reason for this is that data analysis is mainly decentralized to single factory lines and single heavy machines and dependent on employees' own motivations to perform the analyses.

Two of the case firms have established remote connections to the factories and units that allow centralized machine data analyses by their research units located at another location. For example, the product research unit of the pulp and paper company is able to remotely help the factories, even though the experts are scattered around Europe. This allows the company to pool together the data scattered around different factories and deploy centralized support to any site. In addition to intrafirm remote connections, interfirm remote connections are also possible to some extent. Both the chemical and pulp and paper companies we studied allow remote access by some machine suppliers to their production sites to provide technical support.

> To some extent, we have remote connections in place. Sometimes, a supplier takes a remote connection, but it is not constant. If problems occur, we ask certain firms to take a look at them. – *Electricity and Automation Engineer, Chemicals*

While there are already some remote connections, service and maintenance are very much grounded in reactive and time-based practices. While companies collect and send process data to databases for storage and later analysis, the data are seldom used. Proactive maintenance is currently limited to regular maintenance schedules based on time intervals instead of predictive maintenance and machinery conditions being continuously monitored and evaluated. The case firms are not using the available data to develop the service and maintenance process. Even though the interviewees identify the possibility to use the available data to recognize early symptoms of technical failures, the usage of the existing data is mainly triggered by an incident. As an operations manager in the marine company explained:

> If something has happened, like a machine has shut down, then we go through the automation system data to try to figure out what the problem is. – *Operations Manager, Marine*

To summarize, although many technical aspects of IoT (such as sensors, databases, and remote connections) have been implemented, most of the opportunities with the gathered data have yet to be fulfilled. More systematic and full-scale data integration from sensors and machines is lacking.

Value Realization Constraints in Heavy Machinery Companies

Next, we focus on the value realization constraints by companies using heavy machinery. We found three firm-level data-related value realization constraints: (1) organizational value realization constraints, (2) resources and capability value realization constraints, and (3) sensemaking value realization constraints. Our findings are summarized in Table 3.

Although the case firms have invested in the hardware to collect and store data, they have not changed the organization in a way that supports the analysis and use of these data. First, data gathering and analysis are organized within the production lines of factories and with single or few heavy machinery units rather than being collected and analyzed across many production lines and heavy machinery units. The collection and analysis of data are done in silos, and there is no single place to find the data nor ways to explore the data more collectively and systematically. As a quality manager explained to us:

> Organization silos are one obstacle. We have quite limited areas of responsibility, and we do not look at big picture much. The IoT and looking at the big picture are big changes compared to current operations. People have been looking only at their production line and not at neighboring lines. – *Quality Manager, Electrical Equipment*

Another organizational data and IoT value realization constraint that we found was that managers and workers lacked excitement about developing data utilization and the IoT. In part, it was explained to us that this is since the costs and paybacks of investments in the IoT are hard to estimate. Furthermore, there is simply a lack of time between production and maintenance to analyze or develop organizational practices within the single factories or units of heavy machinery.

Therefore, there is clearly a lack of both resources and capabilities to use the data and explore the value in the data. The managers explained that the current workforces at the case firms' production facilities are not equipped with the skills required to utilize the existing data. Furthermore, they are also not incentivized to acquire the needed skills. Finally, managers and workers are not provided with data analysis tools and software to make sense of the data. As an electric maintenance manager explained:

> We use data very poorly. We have the weakness that data are only researched when a failure occurs. Only then do we research the causes but not beforehand. We should be doing this more, and we should have resources so that somebody would research this more deeply. – *Electrical Maintenance Manager, Chemicals*

As such, adding more data may not be a solution before the case firms develop capabilities to apply the existing data and before new tools are in place to utilize new data.

> We already have too much information from the automation system to our operational employees. However, on the other hand, this information could be used better when disturbances occur so that the diagnostics would tell the employees what has happened. – *IT Manager, Chemicals*

Furthermore, the middle managers we interviewed revealed that the focus of top managers has been more on cutting costs than generating new growth avenues. This resulted in both physically limited resources and atmospheres that do not encourage experimentation. Data analysis is considered an additional task on top of daily routines. With limited resources for data analysis within the companies, factories, and units operating heavy machinery, the development of solutions to the current data and IoT implementation challenges have been increasingly moved toward external partners. As one manager we interviewed explained:

> Our maintenance organization is getting slimmer and slimmer, and production organization takes care of production more autonomously, but they do not think that much about the equipment condition […] The slimmer our organizations are getting, the more we need external help. – *Electrical Maintenance Manager, Chemicals*

Finally, our study also points out one central value realization constraint from the data and IoT for middle managers' constrained sensemaking of technological solutions and its value. The interviewed middle managers found it difficult to understand the IoT and digitalization. They were not aware of how they could improve the use of existing data, especially with the current resources. For example, a middle manager we interviewed had a hard time seeing the substantial benefits of knowledge sharing between factories.

> Maybe some special cases that do not occur often would be beneficial to know how they are handled in different factories or about how different processes are carried out in different factories. – *Maintenance Manager, Pulp and Paper*

To conclude, as of now, only a small proportion of the data is used in decision-making in operations and maintenance in firms using heavy machines. This finding is congruent with earlier studies about low data usage (Accenture 2015a; Manyika et al. 2015).

Predictability as the Key IoT Value Driver

Next, we move to our findings regarding how middle managers perceived the potential of the IoT to advance the development of value from heavy machinery data. Our study suggests four key value drivers from IoT in heavy machinery operation: (1) product optimization, (2) maintenance and recovery optimization, (3) energy efficiency, and (4) safety improvements. The four value drivers are all founded on advances that can be made with IoT with respect to predictability. Our findings on the key value drivers are summarized in Table 4.

We find that middle managers perceived *the potential to increase the predictability of production machinery and processes* as the key underlying value driver of the IoT. Predictability encompasses improved information on machine failure symptoms, relationships between the physical conditions and production, and production process lead times. The middle managers observed that increased predictability could create value in multiple ways, such as decreased down times through predictive maintenance, the further optimization of production processes through decreased wait times between production processes, increased energy efficiency, and improved safety.

> We should develop predictability about when our production equipment is about to break down. It would be interesting to know beforehand [...] If we could prolong all equipment maintenance to stoppages without surprises, it could be very valuable [...] Predicting that equipment lasts until the next stoppage would be useful so that we would not need to fix them too early. If that information would be available two to three weeks beforehand, it would be good, since then we would not need to shut down production for separate stoppages. – *Maintenance Manager, Pulp and Paper*

Predictive maintenance would decrease the number of disturbances and increase equipment reliability, factory availability, and output. The interviewees focused more on the potential to increase revenues rather than on cost savings. According to the middle managers, predictability could also enable further production optimization. The case firms have thousands of machines and components in their production facilities, and there is room to improve the understanding of their interrelationships. This would require both a better understanding of their equipment as a system and focused monitoring on the most important machines for the continuation of operations. Furthermore, a more systemic understanding of the production processes and the underlying environmental factors would then help to improve factory equipment life cycle management and would result in cost savings in investments.

> In addition to single components, it is important to understand machine systems and especially critical ones to be able to recover faster from disturbances. All the bottlenecks and all common systems are probably the places where this kind of monitoring comes most. – *Development Manager, Pulp and Paper*

Additionally, all of the focal firms acknowledged the importance of energy efficiency. There is significant potential for improvement. For example, production is currently not yet controlled using energy prices. However, the IoT was not unanimously seen as an answer to improved energy efficiency, even though many of the interviewees believed in this.

> There is a lot of potential in energy efficiency. In that area, we can benefit a lot from IoT systems, and then we could connect different factories. This works definitely. – *Plant Manager, Pulp and Paper*

The middle managers also observed potential to improve safety with better predictability. The interviewees stated that a reliable and predictable factory is also a safe factory. Maintenance work is not carried out in a rush, and operational employees face fewer surprises. Employees can be better prepared for their work in production and maintenance when they know the changes in environmental and machine conditions, which then affect their choice of tools and mind-set. Unprepared and ad hoc activities would then be reduced, which would also decrease the number of safety hazards. A decreased number of accidents would also result in an improved work environment, in fewer sick days, and finally in improved employee efficiency.

> It could be summarized that fewer unexpected problems that need uncertain and quick fixes result in more predictability, so that the problems can be fixed with greater control, and this all leads to higher safety. – *Production Manager, Electrical Equipment*

To conclude, collecting a vast amount of data is not enough to reap the IoT benefits described above. Knowledge sharing within and across production units and across organizational boundaries is needed. Systems for integrating and analyzing data must be in place.

Discussion

Our findings on the present state, constraints, and key value drivers of the IoT from the middle management perspective provide an "organizationally grounded" view of the IoT. While there has been much recent interest in the IoT, our case companies' middle managers possess a more conservative view on the IoT than those previously presented, coined *an evolutionary view on the IoT*. Our findings expose how middle management views IoT implementation as a gradual improvement to current operations rather than a revolution. As such, smart manufacturing, Industrie 4.0, and the IoT as revolutionary, disruptive, and a generator of novel business models is not how middle managers describe and perceive the IoT. Rather, our study points to predictability as being the most central value driver from the viewpoint of middle managers.

Managerial Implications

Our findings suggest that predictability as an underlying key value driver of IoT could be a better conceptualization for joining middle managers, top managers, outside experts, and firms in developing data and IoT systems in heavy machinery. Thus, in order to further implement IoT, top management should utilize the organizationally grounded view of middle managers. This organizationally grounded approach for IoT implementation would allow firms to harness the broader existing expertise from the factory and engage this expertise with the business unit, corporate level, and wider heavy machine IoT ecosystem strategy.

A second managerial implication derived from our findings is that organizations are in need of what we term *connected expertise*. With connected expertise, we mean a pooled group of experts that engage in real-time analysis work and decision-making with continuous information flows from multiple heavy machinery units.

Our results show that current knowledge and data are siloed within factories and units, far from the envisioned connected expertise. This present state should be the starting point for any IoT project. The distributed and disconnected knowledge and data need to be integrated to establish connections between data and experts. Such knowledge and data can be pooled in data and analysis hubs (also termed control or expert centers). Hubs must be able to be easily connected with suppliers. This will require the transformation of organizational practices and the disentanglement and removal of existing barriers between silos. Connected expertise could then gradually arise in hubs, where the best experts from various parts of the organization explore a pool of data in real time as a consorted effort. The role and skill set of the middle manager will need to be accordingly updated with data analytics understanding and adoption of exploratory skills in the same way as air traffic controllers.

The third managerial implication is that use of data and the IoT should be seen as a longer-term adaptation and organizational transformation project. Rather than portraying IoT strategies as system-level rapid implementation projects, the transformation of resources, capabilities, and sensemaking should be seen as programmatic long-term adaptation efforts. A digital modular "architectural view" on the transformation of the organization is needed to allow for the connecting of data from machines, sensors, devices, and humans. The longer-term adaptation program has another systemic-level implication. There are value drivers from the IoT that are clearly more easily implemented and have a high value. However, there are IoT value drivers that are much less clear with respect to payback time and overall value generation potential and are challenging to implement. Thus, systemic elements should also be carefully considered from the start, such as interfaces and platform governance structures, when planning data utilization and IoT project road maps. There is much required groundwork at factories and units using heavy machinery to have them engaged, committed, and informed in this endeavor. The chief digital officers (CDO), nominated recently in many firms, will play a central role in these adaptation and transformation programs as part of wider digital transformation efforts by the company top management teams to "encourage risk taking, foster innovation and develop collaborative work environments" (Kane et al. 2015: 9).

Research Implications

Our findings provide several avenues for future research on value drivers of IoT in the heavy machinery industry. As suppliers and customers become more closely connected through the development of the IoT (Burmeister et al. 2015), there is an increasing need to engage middle managers in developing connected expertise. How to advance such connected expertise, where middle managers' current expertise in heavy machinery is deployed, while their expertise in digital technologies and IoT is not yet at an advanced level, is still an open question. We see many opportunities for future research to study these questions. Based on our findings, we see that researchers could study how companies can refrain from too much top management leading data and IoT development projects and, instead, facilitate the active engagement of middle managers. For example, how could the IoT and data projects be organized, led, and scoped to incorporate the gap between expertise in the IoT and digital technologies and middle managers' expertise? Another question that would be important to study is how to incentivize middle managers to develop value from data and initiate innovation in IoT value creation. For instance, what roles and responsibilities could middle managers be given, and how could their expertise be actively updated and incorporated?

We also see much need for studies on how machinery firms across the value chain are exploring novel innovations and co-creation structures (Gronroos and Voima 2013) in IoT development and implementation with middle managers' active engagement. Insights from successes and failures in novel innovation and co-creation structures with active middle managers are important. This will allow us to take further stock of the opportunities from the IoT that otherwise are constrained by existing siloed organizational and technological structures that do not accommodate data access across machines, functions, and organizations.

Finally, as the IoT integrates supply chain networks where the information flows are easily transferred in networks (Burmeister et al. 2015; Koch et al. 2014), it requires new forms of interfaces and governance structures. While software developers have been critical in much of platform business development, the platforms in heavy machinery rely on the expertise in machines and their operation and management practices in which middle managers have played a critical role. An important question that requires further exploration relates to the novel governance structures and interfaces and how they can bridge the gap between middle managers and software experts.

References

Accenture. (2015a). *The growth game-changer: How the industrial internet of things can drive progress and prosperity*. Retrieved from https://www.accenture.com/_acnmedia/Accenture/Conversion- Assets/DotCom/Documents/Global/PDF/Dualpub_18/Accenture-Industrial-Internet-Things-Growth-Game-Changer.pdf

Accenture. (2015b). *Winning with the industrial internet of things*. Retrieved from https://www.accenture.com/us-en/~/media/Accenture/Conversion-Assets/DotCom/Documents/Global/PDF/Digital_1/Accenture-Industrial-Internetof- Things-Positioning-Paper-Report-2015.pdf

Annunziata, M., & Evans, P. C. (2012). *Industrial internet: Pushing the boundaries of minds and machines*. General Electric Report.

Atzori, L., Iera, A., & Morabito, G. (2010). The internet of things: A survey. *Computer Networks, 54*(15), 2787–2805.

Atzori, L., Iera, A., & Morabito, G. (2017). Understanding the internet of things: Definition, potentials, and societal role of a fast evolving paradigm. *Ad Hoc Networks, 56*, 122–140.

Balogun, J. (2003). From blaming the middle to harnessing its potential: Creating change intermediaries. *British Journal of Management, 14*(1), 69–83.

Bradley, J., Barbier, J., & Handler, D. (2013). *Embracing the internet of everything to capture your share of $14.4 trillion*. White Paper, Cisco.
Burmeister, C., Luettgens, D., & Piller, F. T. (2015). Business model innovation for Industrie 4.0: Why the "industrial internet" mandates a new perspective. *SSRN Electronic Journal, 0*, 1–31. https://doi.org/10.2139/ssrn.2571033.
Gioia, D. A., Corley, K. G., & Hamilton, A. L. (2013). Seeking qualitative rigor in inductive research: Notes on the Gioia methodology. *Organizational Research Methods, 16*(1), 15–31.
Gronroos, C., & Voima, P. (2013). Critical service logic: Making sense of value creation and co-creation. *Journal of the Academy of Marketing Science, 41*(2), 133–150.
Huy, Q. N. (2001). In praise of middle managers. *Harvard Business Review, 79*(8), 72–91.
Jacobides, M. G., Knudsen, T., & Augier, M. (2006). Benefiting from innovation: Value creation, value appropriation and the role of industry architectures. *Research Policy, 35*(8), 1200–1221.
Koch, V., Kuge, S., Geissbauer, R., & Schrauf, S. (2014). Industry 4.0: Opportunities and challenges of the industrial internet. *Strategy & PwC*.
Kane, G. C., Palmer, D., Phillips, A. N., Kiron, D., & Buckley, N. (2015). *Strategy, not technology, drives digital transformation*. MIT Sloan Management Review and Deloitte University Press, 14.
Manyika, J., Chui, M., Bisson, P., Woetzel, J., Dobbs, R., Bughin, J., & Aharon, D. (2015). *Unlocking the potential of the internet of things*. McKinsey Global Institute. Washington, DC.
Porter, M. E., & Heppelmann, J. E. (2014). How smart, connected products are transforming competition. *Harvard Business Review, 92*(11), 64–88.
PwC. (2015). *The smart manufacturing industry: The industrial internet creates new opportunities for Swedish manufacturing companies*. Retrieved from http://www.pwc.se/sv/publikationer/verkstad/the-smart-manufacturingindustry.html
Rouleau, L. (2005). Micro-practices of strategic sensemaking and sensegiving: How middle managers interpret and sell change every day. *Journal of Management Studies, 42*(7), 1413–1441.
Wooldridge, B., Schmid, T., & Floyd, S. W. (2008). The middle management perspective on strategy process: Contributions, synthesis, and future research. *Journal of Management, 34*(6), 1190–1221.

Customer Value in the Sharing Economy Platform: The Airbnb Case

Hong Ngoc Nguyen, Timo Rintamäki, and Hannu Saarijärvi

Abstract The sharing economy is a platform-based business model in which users are directly connected for creation, sharing, and exchange of goods or services that draw on underused resources. While this emerging phenomenon has been studied from several perspectives, including the technical, social, and economic, limited investigation has been conducted from the customer perspective. A study was carried out to narrow this research gap by applying the customer viewpoint to explore and analyse how the sharing economy reconfigures value creation. To reach the goal for this research, an interpretive approach was taken to the case of Airbnb. Customers' experiences of using the sharing economy were examined to disentangle the economic, functional, emotional, and symbolic benefits and sacrifices that together capture the diversity of customer-perceived value of the sharing economy. Customer value is discussed as an important conceptual tool to identify and pinpoint the distinguishing characteristics of the sharing economy and to explore how conventional businesses and emerging sharing-economy platforms can recognise and capitalise on their competitive advantages. While the sharing economy gives customers alternatives that involve easier consumption methods at lower cost, it also provides a unique, personal, and socially integrated experience. This can contribute to values that extend beyond traditional hospitality management and thus imbue sharing-economy platforms with a unique and sustainable competitive advantage.

Keywords Customer value · Sharing economy · Service platform

Introduction

The sharing economy, also known as collaborative consumption, is a peer-to-peer business model built around platform thinking. In this model that has emerged in recent years, peers conduct business activities via a single virtual marketplace

H. N. Nguyen · T. Rintamäki (✉) · H. Saarijärvi
Faculty of Management, University of Tampere, Tampere, Finland
e-mail: timo.rintamaki@staff.uta.fi

© Springer Nature Singapore Pte Ltd. 2018

A. Smedlund et al. (eds.), *Collaborative Value Co-creation in the Platform Economy*, Translational Systems Sciences 11,
https://doi.org/10.1007/978-981-10-8956-5_12

without any other intermediaries. The consumption of goods or services is based on shared access to, rather than exclusive ownership of, resources from the marketplace community (Denning 2014). Adoption of the sharing-economy model has grown significantly in recent years and has become an interesting research phenomenon. Many consider the sharing economy to be the next industrial revolution (Botsman and Rogers 2010) or a 'disruptive business model' (Guttentag 2013), able to redefine industries and consumption. A study of European markets shows that the five key sectors of the sharing economy – collaborative finance, peer-to-peer accommodation, peer-to-peer transportation, on-demand household services and on-demand professional services – facilitated transactions valued at $28 billion in 2015 and projected to reach $570 billion in 2025 (PwC 2016).

Research into the sharing economy has shifted from studying individual sharing platforms to studying the entire sharing economy as a general phenomenon, whatever the product/service types or shared objects may be (e.g. Matzler et al. 2015). Technically oriented work has focused on the sharing platform, how it functions and how humans and technology interact in peer-to-peer networks (Avital et al. 2014; Bucher et al. 2016). Studies taking a social perspective have addressed human motives for sharing (Guttentag et al. 2017; Hamari et al. 2016; Milanova and Maas 2017) the role of trust and reputation (Ert et al. 2016), racial discrimination (Edelman et al. 2017), and impact on labour regulation and society (Kneese et al. 2014; Teubner 2014). Tussyadiah (2015) conducted a study of the sharing economy from a customer satisfaction perspective, to detect the factors that influence customers' choices. Although considerable research has recently been devoted to exploring the sharing economy as a phenomenon, only limited attention (Camilleri and Neuhofer 2017; Tussyadiah and Zach 2017) has been paid to understanding it from a user perspective.

Understanding the research phenomenon from a customer standpoint is important for many reasons. First and foremost, customers are the primary initiators of a sharing economy (Botsman and Rogers 2012). A sharing economy may also have unique characteristics that can lead to customer experiences extending beyond traditional industry logic. Accordingly, studying the sharing economy from the customers' perspective could reveal useful insights that might not only help conventional tourism actors to reconfigure their business concepts but also deepen our understanding of this emerging phenomenon. Although the characteristics of service platforms and the associated management strategies are of importance, losing sight of the customer creates a risk of forgetting the basic nature of the phenomenon, unleashing of unused customer resources for consumer-to-consumer value creation. Not all customer resources are valuable to all customers nor are they necessarily competitive when compared to offerings under traditional business models. Hence, understanding what kind of value is created, for whom, remains a key question for those designing and managing competitive service platforms. Consequently, this paper is devoted to exploring and analysing how the sharing economy reconfigures value creation. Accordingly, the focus here is on identifying the benefits and sacrifices of the sharing economy as perceived by customers, in a tourism context.

We begin by discussing the sharing economy and customer value as key conceptual tools. After a summary of the methodology – including description of the case study research setting – the results related to customer-perceived benefits and sacrifices are reported and discussed. We end the paper with our conclusion and address some limitations and future research directions.

Theoretical Background

The term 'sharing economy' or 'share economy' was first used to describe social welfare in which participants share in pursuit of the greater common good (Weitzman 1986). Since the 2000s, sharing-economy models have attracted the attention of both researchers and practitioners (Smolka and Hienerth 2014). 'Sharing economy', along with alternative terms such as 'collaborative consumption' (Botsman and Rogers 2012), 'consumer sharing system' (Lamberton and Rose 2012), 'access-based consumption' (Bardhi and Eckhardt 2012), and 'the mesh' (Gansky 2010) (cf. Belk 2014), has become a buzzword and is now widely discussed by economists, philosophers, marketers, and entrepreneurs alike (Botsman and Rogers 2012) (Table 1).

According to these definitions, a sharing economy can be perceived as a socio-economic model based on shared usage, promoting temporary access, and functioning as a non-ownership model. Access to, rather than ownership of, products or services is encouraged (Botsman and Rogers 2012). It is important to understand and discuss the when, how, what, and why of sharing, as these offer complementary perspectives on what characterises the very nature of the sharing economy as a research phenomenon. Although this paper focuses on contributing to understanding of why sharing emerges by exploring and analysing the sharing economy from the perspective of the customer (see Fig. 1), other perspectives are discussed in more detail too, for a holistic view of the phenomenon.

Table 1 Selected definitions of 'sharing economy'

Authors	Definitions
Lamberton and Rose (2012, p. 109)	'[M]arketer-managed systems that provide customers with the opportunity to enjoy product benefits without ownership'
Bucher et al. (2016, p. 318)	An economic model in which consumers use online tools to collaborate in owning, renting, sharing, and trading goods and services. A practice enabled and driven by technology
Botsman and Rogers (2012, p. 15)	'Traditional sharing, bartering, lending, trading, renting, gifting, and swapping redefined through technology and peer communities'
Bardhi and Eckhardt (2012, p. 881)	Consumption models in which access is enabled through sharing or pooling of resources/products/services redefined through technology and peer communities
Botsman (2013, p. 6)	'[An] economic model based on sharing underutilized assets from spaces to skills to stuff for monetary or nonmonetary benefits'

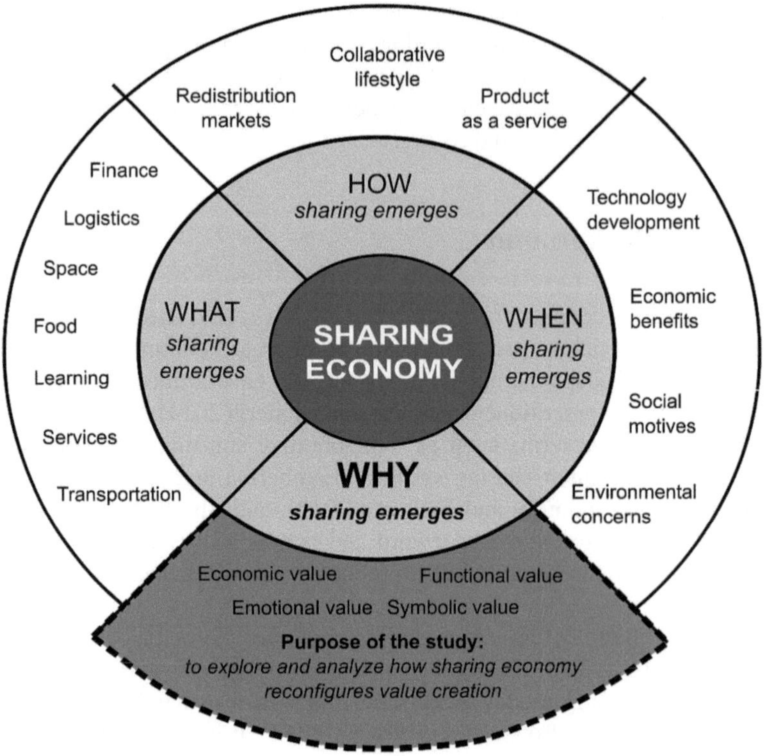

Fig. 1 Summary of the sharing economy and positioning of the study

Why Sharing Emerges: Understanding Customer Value

From a customer perspective, the sharing economy is about new ways of getting value from old resources. Hence, instead of looking only at motives of customers or the attributes of the service platform, we aim to uncover the outcomes of the use of the sharing-economy platform. Understanding these outcomes is at the heart of customer value, which can be defined through benefits (positive consequences of use) and sacrifices (negative consequences of use) (Woodruff 1997; Zeithaml 1988). Customer value is created when benefits are increased and/or sacrifices decreased. The importance of customer value has been widely recognised since the 1990s, and consideration of it has become mainstream in marketing research. However, it is difficult to state a single definition for the concept (Gallarza et al. 2011). Holbrook (1999, p. 5) defines customer value as an 'interactive, relativistic preference experience'. Customer value as a concept is grounded in the idea of interaction between the user and the service. It is an evaluative outcome that is contingent on several contextual factors, among them time, place, and the persons involved. The concept explains behavioural intentions, as perceiving value contributes to preference formation. Rather than offering a narrow perspective of a single stage in a decision process or a single point in time, customer value is judged in evaluation of the outcomes of the entire customer experience. Understanding the characteristics underlying the processes of value creation and perception is perhaps more important for sharing-economy platforms than in many other contexts, because of a 'multisidedness' that often adds complexity. With sharing-economy platforms such as Airbnb, the digital interface steers and (to some extent) standardises interaction, but there is a clear role of direct and uncontrollable interaction also. Tools exist for contextual decision-making, yet there is little control over the actual context during the Airbnb homestay itself. Nonetheless, all these issues and many more potentially affect perceived customer value.

Scholars and practitioners should understand not just whether value is created but also the kind of value that is created and perceived. Viewing customer value from the perspective of dimensions of value provides a deeper understanding of the nature and potential of customer value. In our work, the lens of various dimensions (economic, functional, emotional, and symbolic) serves as conceptual means to address the complexity and diversity of the sharing economy. Shifting attention from a one-dimensional picture to the reality of multi-dimensional customer value helps to bring a richer and thicker understanding of customer perspectives on the sharing economy.

Economic value is defined as a 'product's objective monetary worth to a customer adjusted for the availability of competitive substitute products' (Smith and Nagle 2005, p. 41) or simply as a low price or the best trade-off between quality and price (Gale and Wood 1994; Zeithaml 1998). Economic value is considered to be the 'hard-to-beat' driver of customer value (Rintamäki et al. 2007). In a sharing economy wherein people provide a service from the resources they already have available, they are offered a better price without an addition to standard operation expenses. Those who, in turn, use services from private providers, as in the sharing economy, may find lower-cost options than those offered by conventional providers.

Functional value is defined as 'perceived utility acquired from an alternative's capacity for functional, utilitarian, or physical performance' (Sheth et al. 1991, p. 160). Functional value is focused on solutions that meet a customer's needs with less time, effort, search cost, and decision cost (Rintamäki et al. 2007). A service has functional value when it has sufficient desired characteristics and performs the functions sought (Smith and Colgate 2007). On the functional-value dimension, the sharing economy is considered in terms of the solution it offers customers. Customers' perceptions of the sharing economy's functional value extend to how sharing-economy systems work, how convenient the system is, and whether the new way of consuming helps the customer save both time and energy.

Emotional value is defined as 'perceived utility acquired from an alternative's capacity to arouse feelings or affective states' (Sheth et al. 1991, p. 161). The emphasis is on the feelings derived from the customer's experience (Rintamäki et al. 2007). Value on the emotional dimension can be understood as the evoking of positive feelings (pleasure or enjoyment), social elements (encouragement of bonding, interaction, or trust), or epistemic value (piquing curiosity or increasing knowledge) (Smith and Colgate 2007). In the sharing economy, social interaction between individuals is relatively strong. In this context, users not only consume a service but also engage with other individuals, including the service provider (Lamberton and Rose 2012). The public regard the sharing economy as a new way of doing business, one in which they may have new experiences and feelings when using the service. Attention to the emotional-value dimension could involve analysing customers' feelings and affect, including how they feel about themselves in connection with the service, feelings towards others, and feelings about the sharing economy as a whole.

Symbolic value is defined as positive meanings associated with the consumption that are attached to self and/or communicated to others (Rintamäki et al. 2007; Smith and Colgate 2007). Symbolic value is what the products or services mean to customers or how the item helps customers communicate their identity. For example, some products help customers feel good about themselves in the possessing or giving. Other products are valued because of the personal meanings tied to them. Some products or services help customers to express their personality, tastes, status, or chosen image to others (Smith and Colgate 2007).

Although customers may justify their choices in terms of functionality or utility, it is often the symbolic value that underlies these choices (Sheth et al. 1991). In the case of Airbnb, customers may opt for collaborative consumption because of anti-consumption spirit or a wish to lead an ecologically friendly lifestyle (Lamberton and Rose 2012). Customers with environmental concerns may perceive a collaborative lifestyle as an ideal solution for protecting the environment: participants may use fewer resources and produce less waste. The anti-consumption-minded customers value the psychological gains of choosing an alternative way of life that entails consuming less and sharing more (Lamberton and Rose 2012). By considering the symbolic dimension, one can gain understanding of a type of value customers commonly find in the sharing economy: what the item means to them or how they use it, which reflects the customer's personality and ideal of self-reliance.

When the Business Potential Emerges: The Drivers of the Sharing Economy

Four main sets of drivers exist that collectively contribute to the development of the sharing economy. Firstly, advances in technology, especially tied to the boom in Internet usage, have radically changed the marketplace. Business structures must be reshaped to adapt to new customer habits and interests, and simultaneously customers are adopting a new way of communicating, perceiving, and consuming – using new forms of social networks. This change in customer behaviour is characteristic of the 'digital era' (Denning 2014; John 2012, 2013). Emerging from the digital market, the sharing economy is enhanced by multiple technology drivers. New technologies have equipped sharing systems with the necessary infrastructure to make the searching and sharing process more convenient and efficient. New systems such as online payment, GPS-based navigation, and mobile devices are critical tools contributing to the development of the sharing economy (Botsman and Rogers 2012). Hence, technological developments are a potential source of functional value.

Secondly, the sharing economy yields economic benefits for both providers and consumers. For the provider, new sources of income from an existing resource are unlocked: a car or flat can become an avenue for individuals to earn money. Through a collaborative-consumption platform, people can readily utilise personal resources to start business operations, with collaborative platforms providing a mechanism to manage marketing-related tasks. The individual can leverage the infrastructure of a large platform of this type to market products or services (Botsman and Rogers 2012). A recent US government report highlighted the financial benefits that arise from utilising the marketing power of such a platform (Budget and Legislative Analyst's Office 2015). Thanks to the direct connection of buyers to sellers, a competitive price point can be established. Though the providers lack such advantages as volume benefits, the direct connection minimises middleman margins and supply-chain markup (Denning 2014). The collaborative process centred on the consumption transaction is made more efficient and convenient via the support of advanced technology such as systems that match a consumer's criteria with provider offers, smart devices, and technology-assisted navigation tools. Mediation by technology affords, rather than detracts from, the access that collaborative-consumption customers gain to better products at lower prices, simpler transaction processing, and savings on time and effort (Denning 2014; Smolka and Hienerth 2014).

Next, from a social point of view, there may be several factors in the dramatic development of the sharing economy. Sharing is a core impulse of human beings (Smolka and Hienerth 2014). One witnesses this when people share books at a public library or share space in a public park. Similarly, social networks' users share ideas, information, and knowledge via personal blogs. Researchers consider sharing to be a voluntary act motivated by desire to diffuse resources to the community (Botsman 2014; Dahlander and Magnusson 2005) or a response to extrinsic or intrinsic motivation to participate in the community (Franke and Shah 2003) and contribute to the public good (Lerner and Tirole 2002). Recently, social networks such as Facebook, LinkedIn, and YouTube have been nurturing users' habits of sharing. They adopt this sharing pattern across an expanded circle of people. While familiar circles such as relatives, co-workers, and colleagues connect users, the circle has been extended to include users around the world with common interests. Online, many people now build their social profiles by sharing pictures, contacts, history, reviews, comments, and votes. The act of sharing has become common practice in social networks (Garbarino and Strahilevitz 2004). Those who adopt sharing habits online are widely connected and enjoy a sense of mutual trust.

Finally, collaborative consumption provides sustainable solutions that resolve the environmental concerns of many (Firnkorn and Müller 2011; Truffer 2003). In the twenty-first century, growth of the production and marketing industry has contributed to a consumption society. New products and services are continuously developed to meet customers' demands as those wants and needs diversify further and keep changing. Hence, goods are quickly created, are soon used, and swiftly get discarded. In this climate, the life cycle of products is growing shorter and amounts of waste are rising (Botsman and Rogers 2012). The sharing economy encourages sharing and co-operation, lower quantities of consumed material, and more accessibility (Bardhi and Eckhardt 2012). In addition, collaborative consumption enhances an environment-friendly lifestyle of using less material and producing less waste (Botsman and Rogers 2012; Lahti and Selosmaa 2013). The sharing economy is attractive partly because of its sustainable consumption and smaller environmental impact (Piscicelli et al. 2015).

How the System Works: Characteristics of a Sharing-Economy Platform

There are various sharing systems through which resources, capacity, and processes are delivered between peers. First, product as a service systems emphasise the effectiveness of reusing products to satisfy users' needs (Botsman and Rogers 2012). Some of these systems are based on usage rather than ownership, while others offer a change of ownership aimed at use and extension of its life. Second is redistribution markets, in which various objects offered online are shared through peer-to-peer networks (Botsman and Rogers 2012). The network's users upload descriptions of items such as their old computers or clothes to a redistribution website and then share these items with people indicating a need for them. On these sites, products are exchanged at no charge, for cash, for points, or in some combination of these methods. Users can even barter items for similar products of the same value. Transactions are based on three key principles: reciprocity, fairness, and use of a review system. In the third system type, the sharing economy offers a common platform to bridge different needs. Users with similar interests engage and share their time, space, skills, ideas, or money. These exchanges nurture a new commons, the 'collaborative lifestyle' (Botsman and Rogers 2012). Of whatever type, this lifestyle satisfies the need for connection, mediates the market, and provides an example that others online can reproduce in day-to-day life. Participants combine their individual strengths to create better results for everyone. Collaborative-lifestyle-oriented users with the same needs gather in a common workplace to work together without being involved in the same projects. The shared work environment can improve sociability and generate more interaction between people. Collaborative lifestyles also help to bypass intermediaries who in other markets act to connect buyers and sellers. Banking systems, for instance, serve as a trusted party connecting people who want to borrow money with those who want to lend it. The 'money-renting' services have now rebranded themselves as money-lending systems without banks as intermediaries (Funk et al. 2011). The connection occurs sooner, and the cost of borrowing money is lower.

What Is Being Shared: The Objects of Collaborative Consumption

Today, sharing economy can be found in industries that run the gamut of finance (money lending, crowdfunding, and cryptocurrencies), products (pre-owned goods and rental goods), space (work space and accommodation), transport (car sharing and driver optimisation), logistics (local delivery, shipping, and storage), services (professional and personal services), food (shared food), and (instructor-led and peer-to-peer) learning. When there is a need, the sharing economy gives customers an alternative way of consuming, from renting to swapping or sharing, instead of buying a new product or service (Botsman and Rogers 2012). In consequence, companies such as Kickstarter, Craigslist, Uber, LendingClub, and Airbnb – to name but a few – have established themselves at the forefront of this business transformation (Owyang 2016). In increasing numbers, researchers are suggesting that traditional organisations should prepare for further growth of the sharing economy (Botsman 2014; Botsman and Rogers 2010; Cusumano 2014; Denning 2014).

Research Methodology

Given its purpose of exploring and analysing how the sharing economy reconfigures value creation, our study addressed this phenomenon itself, rather its justification (Yadav 2010). This orientation for the research is reflected in the methodological choices taken: we used qualitative methods in general and a case study research setting in particular to explore the perspectives of customers and uncover their experience of the sharing economy in using Airbnb. Delving into the benefits and sacrifices perceived by customers helped us to uncover the themes that reflect the dimensions of customer value. Qualitative methods help to deepen and sharpen our understanding of the social world, and a case study is 'an empirical inquiry that investigates a contemporary phenomenon in-depth and within its real-life context' (Yin 1999, p. 18). The case study method is widely used by disciplines from psychology, sociology, and political science to economics and management (Yin 1994). A case study aids in understanding the phenomenon in its proper context (Dubois and Gadde 2014), which is especially important for fulfilling the aim for our research: understanding the context is critical to uncovering the nature and richness of perceived customer value.

Description of the Case

We chose the online accommodation booking service Airbnb as our case for studying customer value in the sharing economy. This San Francisco-based company was founded as a start-up in 2008 by Joe Gebbia, Brian Chesky, and Nathan Blecharczyk.

In its community, users can rent out or book accommodation. The company offers an approach to bed-and-breakfast accommodation that is built on the collaborative-consumption model (Guttentag 2013). Described more precisely, Airbnb is a peer-to-peer accommodation-renting community that caters to hosts and travellers. Hosts use Airbnb to promote their underused space and rent it to others, while travellers use the site to book a stay at another person's house. Airbnb acts as a third party between hosts and travellers, charging fees to both parties. In 2015, the average Airbnb host in San Francisco earned $440 from renting out spaces via the Airbnb website, with earnings in some cases reaching $1900 per month (Budget and Legislative Analyst's Office 2015). A report released by Airbnb in 2012 stated that 56 per cent of hosts used income from Airbnb to make their monthly house payments or pay rent, while 42 per cent used this money for day-to-day expenses (Airbnb 2012). Airbnb offers three types of room: shared rooms, an arrangement in which guests share the entire space with the host or others and do not have a room to themselves; private rooms, with guests sharing some common areas with their hosts, such as the kitchen, lounge, and bathroom, but having their own bedroom; and entire homes/flats, a whole unit rented by the guests, who need not share the space with the host or with anyone else. The type of accommodation ranges from one- to three-bedroom flats to one-bedroom studio flats with a kitchen or even a treehouse bungalow.

Airbnb serves as an interesting and representative case of the sharing economy for three main reasons. Firstly, it applies a typical collaborative-consumption business model utilising critical mass, idle capacity, belief in the common good, and trust between strangers. Airbnb has critical mass as a widely available service with booking available in 190 countries and more than two million listings at any given time, worldwide (Airbnb 2017). Idle capacity is Airbnb's speciality: the company enables customers to rent out unused rooms via its service. Airbnb customers, in turn, gain trust in the community by checking online profiles and previous reviews. Secondly, Airbnb, alongside other modern companies, such as TaskRabbit and ZipCar, is a pioneer of the sharing economy and is still evolving. Airbnb's business model has had sufficient time for adjustment and development and is a proven success, accounting for a large slice of the market. In 2012, rooms available for rent via Airbnb exceeded the volume of any single brand used by the two largest hotel chains in the world (InterContinental Hotels Group and Marriott International) (Guttentag 2013). Thirdly, approximately, 60 million guests have used the Airbnb service, in 34,000 cities (Airbnb 2017). Consequently, customers are generally well aware of the characteristics and value-creation potential of Airbnb.

Data Generation and Analysis

For the selection of interviewees, we applied two main criteria. Firstly, participants had to be customers of Airbnb; that is, they were required to have booked accommodation via Airbnb. Secondly, the respondent sample was selected to cover customers of diverse nationalities. This purposive sampling was used to provide a complete picture (Patton 1990) of the data represented and the research phenomenon. A snowball sampling technique was used to identify key informants (Moriarty 1983). We asked the respondents to ask their friends and their friends' friends whether they would be interested in participating in an interview.

We followed a semi-structured interview method, as widely used in social research. An interview protocol was developed that included the conversation flow, important themes, and key general techniques. There was no set order for the conversation, and respondents were allowed to expand on their thoughts, illustrate them, and digress freely throughout (Alam 2005). The interview protocol was used as a guide to ensure that the interviews were conducted in an appropriate and consistent manner. The respondents shared their experiences in their own words and discussed themes that they found significant. Questions were based on the interviewees' responses so as to avoid imposition of any predefined framework or logic on respondents (Alam 2005). If an interesting point was raised during the interview, the interviewee was asked for further explanation. Then, the interviewer verified the explanation by repeating it and asked for clarification, to ensure correct understanding of the subject's responses. Notes were taken during the interviews, and these were compared and reflected on later, in the analysis stage.

In all, 21 semi-structured interviews were conducted, over the span of 1 month (see Table 2). Among the respondents were both males and females of various nationalities: British, Chinese, Finnish, Hungarian, Korean, American, and Vietnamese. Ages ranged from 22 to 32 years. The Airbnb spaces were in Europe, North America, Australia, and Asia. Interviews were done as face-to-face meetings, online video calls, or (for subjects who were in other countries) online text-based chat.

Data analysis took place in three, complementary phases. In the first, all interview recordings were transcribed, for 81 pages of text in total (including chat logs). Transcripts were read twice, for an overall idea of the data in general and the customer-perceived value of Airbnb in particular. Secondly, more detailed analysis of the transcripts was conducted. This was considered critically important, as multiple readings of transcripts were necessary for capturing the experience of each informant holistically (Flint and Woodruff 2001). The second phase included thematic analysis for preliminary identification of the themes that emerged from the interviewees' words and to capture the perceived value of using Airbnb (for a similar thematic analysis, see Camilleri and Neuhofer 2017). In the final stage, the themes identified were considered in terms of the economic, functional, emotional, and symbolic dimensions of customer value. Here, focus was placed on systematic

Table 2 The informants

No.	Pseudonym	Gender and age	Airbnb use region	Times Airbnb was used	Interview date	Interview length	Interview method
1	Carol	F, 22	Europe	4	14.10.2015	34 min	Video call
2	Nancy	F, 22	Europe	10	25.10.2015	43 min	Video call
3	Lily	F, 25	North America	6	26.10.2015	36 min	Video call
4	Tomas	M, 32	Europe	2	27.10.2015	56 min	Face-to-face talk
5	Anna	F, 27	Asia	3	30.10.2015	50 min	Online chat
6	Josh and Lyna	M, 22; F, 22	Europe	1	31.10.2015	26 min	Face-to-face talk
7	Hana	F, 28	Europe	3	31.10.2015	19 min	Video call
8	Violet	F, 27	Asia	2	31.10.2015	32 min	Video call
9	Emily	F, 25	Australia	1	1.11.2015	56 min	Video call
10	Lara	F, 24	Europe	7	3.11.2015	51 min	Online chat
11	Tiina	F, 22	Europe	4	4.11.2015	49 min	Video call
12	Linda	F, 25	North America	4	4.11.2015	42 min	Video call
13	Peter	M, 25	Asia	2	7.11.2015	30 min	Video call
14	Teehee	F, 23	Europe	8	7.11.2015	56 min	Online chat
15	Sara	F, 26	North America	1	4.11.2015	16 min	Video call
16	Bella	F, 24	North America	4	9.11.2015	45 min	Online chat
17	Cathy	F, 27	Europe	2	10.11.2015	45 min	Video call
18	Kimmy	F, 22	Europe	1	12.11.2015	31 min	Video call
19	Jonas and Daisy	M, 26; F, 22	Europe	1	13.11.2015	48 min	Face-to-face talk
20	Cindy	F, 25	Europe	1	20.11.2015	51 min	Face-to-face talk
21	Julia	F, 27	Europe	8	26.11.2015	21 min	Video call

and iterative reflection on the kind of value that interviewees perceived when using Airbnb; other opinions or preferences were excluded from consideration (Stake 2004).

Results

Proceeding from the analysis of interview content, we identified six types of benefits and five kinds of sacrifices as underpinnings for the value creation of the Airbnb user experience. To obtain a better understanding of the value-creation process, we then used the conceptualisation of value along the economic, functional, emotional, and symbolic dimensions (Rintamäki et al. 2007) to classify the benefits and sacrifices into more manageable sets of perceptions. These dimensions are summarised in Table 3, with illustrative interview material.

The perceived benefits (monetary savings, convenience, a sense of being at home, experiencing authenticity, social engagement, and enabling of meaningful lifestyles) and sacrifices (risk of unreliable information, time and effort costs, feelings of insecurity, stress, and embarrassment) represent the customers' perceptions of the sharing economy and capture how the sharing economy reconfigures value creation. Alongside the respective dimensions of value, they are discussed below in more detail.

Discussion

We have identified and delved into the multi-dimensionality of the sharing economy as revealed by experiences of Airbnb. These findings can be considered in conjunction with the typology used by Guttentag et al. (2017). In the latter study, respondents were clustered into money-savers, home-seekers, collaborative consumers, pragmatic novelty-seekers, and interactive novelty-seekers. While these clusters reveal various motivations for using Airbnb, they do not extensively address the perceived sacrifices. To this end, we looked at dimensions of value, to go beyond motivation-based segmentation by addressing both the benefits and the sacrifices that using Airbnb may entail.

Economic value is perceived through the monetary savings that are a fundamental benefit of the sharing economy. In a finding consistent with previous studies (Denning 2014; Smolka and Hienerth 2014), interviewees chose to use the sharing economy because it helps them save on costs. They experience economic benefits when they can take advantage of a similar offer at a lower price or of a better offer at the same price (Zeithaml 1998). As was illustrated by the extracts in Table 3, economic value was perceived primarily when the interviewees saw more options to choose from, at prices more competitive than those offered by conventional providers. The services that hosts provide are free of frills; the primary aim is to satisfy basic needs rather than deliver premium service quality. On top of this, the sharing economy can offer additional personal benefits to customers at prices equal to or lower than traditional services. For example, hosts can offer services such as airport pickup or dinners cooked in the local style. At base, this sharing economy offers savings on accommodation expenses, which translate to perceived economic value for its users.

Functional value is increased by one benefit (convenience) and decreased by two types of sacrifices (the risk of unreliable information and the required time and effort). Convenience stems especially from the booking process and terms of accommodation. Advanced technology makes it easy for customers to compare alternatives and find the offers that best match their preferences. The peer-to-peer relationships

Table 3 Summary of customer value, with illustrative quotes

Dimensions of customer value	Benefits and sacrifices	Illustrative verbatim comments
Economic benefits	**Monetary savings**: Airbnb is perceived as lower cost than similar accommodation. Customers enjoy better services or facilities at the same price	'I can save a lot of money' – **Cindy** 'You have the same services [as] in the hotel [but] at lower price' – **Jonas and Daisy** 'At regular hotels, it is more expensive, because we have to pay for other services – such as [a] cleaning fee, service fee, … At Airbnb, we only pay the room price for the owner' – **Josh and Lyna**
Functional benefits	**Convenience** (of the booking process and terms of accommodation): Airbnb helps to simplify the booking process. Customers can negotiate terms and conditions with individual hosts	'It is easy for me to book a room. I can use my Facebook account to sign in very quickly. The user interface is simple' – **Peter** 'I have the application on the phone. I can just choose the date and the city, and I can see the map, I can see the price, all the things directly, and the date available [and] then I can communicate with them. If it's fine, then I just book directly… The payment is very convenient' – **Lara** 'The host helps me with many things, I hardly [ever] get lost when I use Airbnb' – **Hana**
Functional sacrifices	**A risk of unreliable information**: The online information on Airbnb does not reflect actual room conditions. Customers face miscommunication and find it difficult to evaluate the spaces offered	'The room is written as "central of Milano", but actually it is the central of [an]other town, around 30 minutes by train from Milan. I feel a bit cheated… It costs us time and money to travel to and from. I found it so inconvenient' – **Tiina** 'Some places are reviewed nicely, but they could turn out [to be] very bad. It is not trustworthy' – **Violet** 'There was kind of a mix-up there, because the price he quoted on Airbnb was per person although it said different[ly], so we kind of like paid double' – **Tomas** 'We travelled during peak season. Hosts charged us double the quoted price. The room we stayed in was terrible, unlike the room in [the Airbnb] pictures. It did not look like a normal room at all but like a storage/reused room' – **Hana**
	Time and effort: Customers have to deal directly with non-professional hosts and non-standard service quality	'It's really hard to communicate with hosts' – **Linda** 'I had to ask 7–8 places before I got one. It really takes time' – **Lily** 'You have to arrive at a certain time, co-ordinate with the host to have the key, and everything, because they had their own schedule – not always available like hotel reception… Sometimes it is not easy to schedule your time of arrival' – **Julia** 'I like to come on holiday [to[a hotel because you can have someone if you want your towel washed or something. Someone will come, someone will clean, but you don't have that in Airbnb' – **Cindy**

Emotional benefits	**A feeling of being at home**: Airbnb locations are real homes with furniture and home appliances. Customers can follow their established routine and enjoy the premises as they would their own home	'I want to stay at [an] Airbnb place because I like its homey atmosphere' – **Tiina** 'You just like staying [at] your home so. Basically, what we did: we cooked dinner, watched some TV, and sat on the sofa' – **Cindy**
	Experiencing authenticity: Customers can experience local living and culture, from the room decoration to living styles, from food to interacting with local people	'At a local stay, I could experience their lifestyle more. I think that is the point of traveling: for a truly unforgettable experience' – **Bella** 'Cause it's not possible if u are just at a hostel, surrounded by other travellers. If you are at a hostel, people may know "[a] little more" than you but still less than locals know. Even though [a] hostel organizes some "local trip" like pub crawl it's another sightseeing tour. And personally, I value unique experiences – something unlike what others have heard of' – **Teehee** 'You could have a unique experience unlike anything before' – **Anna** 'Host invited me to a local festival. It was an unexpected experience. As if I was coming back to a village and meet[ing] elderly [people] who were friendly and warmly hold you in their arms. I don't think that I am in a foreign country, I feel like I am in my hometown' – **Kimmy**
	Social engagement: The Airbnb mechanism builds trust and encourages kindness between strangers. Customers can make new friends and engage with the community	"I have met the host lady who is really nice. We talk a lot, like Yoga and everything, and when I live there, we cook together. For me, it is nice to meet new people and to stay at someone's home makes me feel really comfortable like a home [and] not a hotel or something else' – **Lara** 'I even get to know some people in the village. They invited us to their social circle because they know my host' – **Tomas** 'When we were in Rome, the host left us cake and wine with a short note: "Merry Christmas." It was Christmastime. It was so nice. So when we left, we gave her some wine and some presents in return' – **Tiina**

(continued)

Table 3 (continued)

Dimensions of customer value	Benefits and sacrifices	Illustrative verbatim comments
Emotional sacrifices	**Feelings of insecurity** (worries about safety and security with Airbnb): Customers stay in a stranger's house without the strict safety and security regulations a conventional hotel or hostel must follow	'Staying at a stranger's place who knows what's gonna happen' – **Bella** 'I like the concept, but sometimes I feel not secure… especially when I was traveling alone so I looked for a woman host…well cause there might be some sexual harassment' – **Teehee** 'The main door wasn't closed. He said he left the door open so that the air could circulate around; otherwise the room will be really hot. I was scared [about] if someone came in when we did not notice' – **Linda**
	Stress: One must resolve issues, without any responsible party to count on	'We could not get the key to enter. The host blamed another guest for losing the key but did not feel sorry… Airbnb didn't play any role in this situation apart from trying to contact her, like us' – **Linda** 'One time I was uncomfortable with Airbnb is that the host in Lisbon was on vacation when we arrived. I messaged him, but he did not reply. When I arrived in Lisbon at 8 o'clock, he was in another city, and he had to drive back. It took 4 hours for him to be back. We had to wait the whole morning, and we could do nothing' – **Hana**
Symbolic benefits	**Enabling of meaningful lifestyles**: One can care for the environment, express a preference for freedom and equality, and role play for out-of-the-ordinary lifestyles	'There was one in particular in London. We stay in a very nice place in [the] London area named Hampstead. Like it [is] kind of the special place in London, very expensive and a lot of celebrities live there. Kind of place I could never live because I [could] never have enough money to afford to live there. So it is really cool. I got [to] stay there via Airbnb' – **Julia** 'At a hotel, there are things like shampoo and stuff, the towel washed every day like crazy. It is quite wasteful. I do not need my towel washed every day […]. So I am interested in the environment and [that] kind of stuff, so maybe [Airbnb is] more environmental' – **Cindy** 'You can step out [of your] comfort zone' – **Kimmy** 'It is more equal. No-one has to serve anyone' – **Tomas** 'When I stay in Kyoto, the room is exactly like a Nobita room. A typical traditional Japanese room: with bamboo mattress, a small table in the middle, near a closet like the one of Doraemon. It was interesting – like I can live the life of my favourite story from childhood' – **Peter**
Symbolic sacrifices	**Embarrassment**: Peer-to-peer interaction results in unique experiences and strong bonds but, conversely, disturbs the privacy of customers	'My partner is a woman…we had to say in advance because we do not want last-minute cancel… with a hotel, we do not think we have to say that' – **Cindy** 'Host is a guy and he is extremely passionate. He comes really close and asks us for party or something. He always touches us, not in [a] weird way but still like for strangers it's a bit weird' – **Lara** 'The lady kept asking me: 'How long have you been with your boyfriend'? 'Why don't you stay at your home'?… I feel like I['m allowed to] stay in someone['s] house as a favour, even though I have paid for It' – **Violet**

of Airbnb mean that hosts can be helpful supporters in the event that customers need assistance. Arrangements between users and providers can be flexible because their relationship is not merely one of buyers and sellers; it is based more on trust and friendship between partners. Customers can negotiate the terms of use and avoid various formal obligations, such as contracts and deposits. However, two kinds of sacrifices may decrease the functional value created. There is a risk of unreliable information: the details online may not accurately reflect actual conditions, as with hosts misrepresenting the quality or location of the accommodation. In addition, customers might have to deal with miscommunication and may find it difficult to find the information that would enable them to evaluate the accommodation. Time and effort sacrifices can result from having to deal directly with non-professional hosts and non-standard service processes. As was well documented in the interviews, long and unexpected waiting times and changes to travel itineraries can all contribute to functional sacrifices with Airbnb.

Emotional value is the sum of three types of benefits (the feeling of being at home, experiencing authenticity, and social engagement) and two types of sacrifices (feelings of insecurity and stress). Customers may experience a sense of being at home if they perceive the accommodation as comfortable and resembling their own home even though they are in another place. The experience of authenticity, in turn, involves the possibility of experiencing local culture in ways that go beyond conventional tourism services. Customers can also feel a closer bond with the service providers (hosts) because the interaction is within a peer-to-peer context. A peer-to-peer relationship is developed with feelings of trust and equality, of being friends rather than business partners. Choosing the sharing economy can encourage customers to interact with providers socially and engage with the community more than would otherwise be possible. Previous studies have suggested that participation in the sharing economy can have the benefit of satisfying basic human social instincts (Franke and Shah 2003; Lerner and Tirole 2002; Smolka and Hienerth 2014). Similarly, the social interaction encouraged by the sharing economy is a development of the interaction experienced in a social network (Piscicelli et al. 2015). The converse of this was seen in feelings of insecurity and stress – sacrifices emerging from anxiety related to unfamiliar people, places, and conditions, along with the stress of having to deal with unpleasant situations. For example, a customer could perceive some hosts as too intrusive, resulting in stressful experiences.

Finally, on the symbolic value dimension, there is one benefit (enabling of meaningful lifestyles) and one sacrifice (embarrassment). In general, symbolic value is bound up with the meanings embedded in the object that customers want to attach to themselves or communicate to others (Rintamäki et al. 2007; Sheth et al. 1991). Some respondents found the possibilities opened by Airbnb and the underlying concept of the sharing economy especially meaningful. For instance, they saw the sharing economy as a way to reflect a new lifestyle. One of the informants said that Airbnb allowed her to live someone else's life in a posh neighbourhood. Customers also are able to use the sharing economy to reflect their personality and as a way to communicate it to others. Using modern collaborative-consumption initiatives instead of conventional service providers allows people to express themselves as independent customers who favour a small carbon footprint or enjoy new, authentic experiences. Consumers' ability to behave in accordance with their environmental consciousness has been identified as one of the benefits most readily available from the sharing economy (Bardhi and Eckhardt 2012; Katzev 2003). In the case of Airbnb, customers prefer the sharing economy on the symbolic dimension for the positive environmental effect associated with utilising unused rooms, minimising unnecessary costs, and saving resources. Conversely, the customer-identified drawback of embarrassment emerged as a potential source of symbolic sacrifice. Although peer-to-peer interaction between customers and hosts can result in social bonding and unique, pleasant, and positive experiences, it can also negatively influence customer' privacy.

This discussion and the customer comments in Table 3 reveal the diversity of perceived value connected with the sharing economy as evidenced in the case of Airbnb. These not only articulate the benefits emerging from use of the sharing economy but also identify the sacrifices that help to shape the overall perception of value. Together, these benefits and sacrifices illustrate both the characteristics and the potential of the sharing economy. From a theoretical standpoint, our results also bridge two scholarly perspectives on the sharing-economy platform, that of companies and that of customers. Customer value is at the core of strategic management, for it explains why customers buy and where they do so. Once we understand the role of the benefits and sacrifices characterising sharing-economy platforms, the strengths and weaknesses of those platforms from a strategic angle will become more evident.

Managerial Implications

For existing companies that have established business models based on the sharing-economy ethos, the benefits perceived by customers serve as a valid starting point for strategy planning. Companies should focus on strengthening and communicating the benefits that fundamentally characterise the sharing economy. These can include monetary savings, convenience, experience, and engagement for customers. Conventional providers may be able to replicate the combination of reasonable prices and convenience, but building the levels of authenticity and social engagement that customers experience with the sharing economy can be much more difficult. However, traditional companies could more clearly emphasise that their system is free of many of the sacrifices that often characterise a sharing economy. For example, additional focus could be put on the fact that customers rarely experience insecurity, embarrassment, and stress with conventional hotels. Awareness of the benefits and sacrifices that characterise both traditional and sharing-economy business models is a promising starting point for further development.

The theoretical approach of considering costs and benefits in combination with the dimensions of customer value confers better understanding of customer behaviour and reveals the reasons behind user choices. From the managerial point of view, firms should focus on either increasing benefits or reducing sacrifices (Zeithaml 1998). They should aim to build high value on at least one dimension or a unique combination along certain dimensions that their competitors would find it difficult to imitate (Rintamäki and Kirves 2017). This addressing of the relevant dimension(s) of customer value, with how well it yields benefits and reduces sacrifices, should then be communicated to both customers and personnel in the form of a customer value proposition. The social, environmental, and non-commercial characteristics of the sharing economy and its perceived value on specific dimensions can create an experience that is beyond the reach of conventional business models.

Limitations of the Study and Avenues for Future Research

As every study does, our research has limitations that should be taken into account. Firstly, although the number of informants was considered adequate for a qualitative study of this nature, the interviewer-interviewee interaction did not always take the form of face-to-face conversational dialogue. The video calls and online chat used to generate some of the data may not have achieved the richness of data possible with traditional interviews. Secondly, the informant demographics were heavily skewed towards younger customers. Therefore, the results capture especially how younger customers perceive the sharing economy; middle-aged and pension-age participants might have emphasised other benefits and sacrifices, while there could also have been significant overlap. Nevertheless, the contribution of approaching the sharing economy from the perspective of customer value's dimensions and in terms of benefits and sacrifices remains important. Finally, while qualitative enquiry yields critically important insight into an emerging phenomenon, a more quantitatively oriented approach could complement this by contributing to a better understanding of the dynamics and causal relations at play in the sharing economy. In this respect, defining new segmentation criteria for the sharing economy (see Guttentag et al. 2017), linking perceived benefits and sacrifices with the sharing economy to key corporate-performance indicators (e.g., Rintamäki and Kirves 2017), and identifying how the 'dark side' of the sharing economy affects repurchase intentions (see Malhotra and Van Alstyne 2014; Tussyadiah 2015) offer interesting avenues for future research.

Conclusion

The sharing economy, as an increasingly prominent and disruptive business model, has become a well-established but evolving phenomenon. Today, it accounts for tens of billions of dollars in revenue per year and has expanded its reach to industries such as finance, logistics, learning, and many service-based industries. The study we carried out to address this phenomenon considered the sharing economy by examining both what it is and why, when, and how sharing emerges. The vantage point we thereby gained provided complementary perspectives on the phenomenon and enriched conceptualisation of the sharing economy. Secondly, to understand why sharing emerges, we devoted attention to exploring and analysing how the sharing economy reconfigures value creation, with the dimensions of customer value chosen as a theoretical tool to address both the benefits and the sacrifices related to the sharing economy. We found that monetary savings, convenience, feelings of being at home, experiences of authenticity, social engagement, and enabling of meaningful lifestyles characterised the benefits related to the sharing economy. Conversely, risk of unreliable information, costs in time and effort, feelings of insecurity, stress, and embarrassment represented the sacrifices customers perceived as accompanying the sharing economy. Taken together, these benefits and sacrifices represent a combination along the economic, functional, emotional, and symbolic dimensions of value that may result in a unique customer experience that is beyond what conventional companies can achieve. Accordingly, it is all the more important to understand the dimensions accorded weight by customers and the benefits and sacrifices that serve as a critically important starting point for any strategic endeavour, whether within or outside the sharing economy.

References

Airbnb. (2012). *San Francisco impact report*. http://blog.airbnb.com/economic-impact-airbnb/?_ga=1.80948052.216156446.1432016310#san-francisco. Accessed 1 Apr 2017.

Airbnb. (2017). *Airbnb, Inc*. https://www.airbnb.com/. Accessed 1 Apr 2017.

Alam, I. (2005). Fieldwork and data collection in qualitative marketing research. *Qualitative Market Research: An International Journal, 8*, 97–112.

Avital, M., Andersson, M., Nickerson, J., Sundararajan, A., Van Alstyne, M., & Verhoeven, D. (2014). The collaborative economy: A disruptive innovation or much ado about nothing? In *Thirty-Fifth International Conference on Information Systems* (pp. 1–8), Auckland.

Bardhi, F., & Eckhardt, G. M. (2012). Access-based consumption: The case of car sharing. *Journal of Consumer Research, 39*, 881–898.

Belk, R. (2014). You are what you can access: Sharing and collaborative consumption online. *Journal of Business Research, 67*(8), 1595–1600.

Botsman, R. (2013). *The sharing economy lacks a shared definition*. Retrieved October 1, 2016, from http://www.slideshare.net/CollabLab/shared-def-pptf.

Botsman, R. (2014). Sharing's not just for start-ups. *Harvard Business Review, 92*(9), 23–25.

Botsman, R., & Rogers, R. (2010). Beyond Zipcar: Collaborative consumption. *Harvard Business Review, 88*(10), 30.

Botsman, R., & Rogers, R. (2012). *What's mine is yours: How collaborative consumption is changing the way we live*. London: Collins.

Bucher, E., Fieseler, C., & Lutz, C. (2016). What's mine is yours (for a nominal fee) – Exploring the spectrum of utilitarian to altruistic motives for internet-mediated sharing. *Computers in Human Behavior, 62*, 316–326.

Budget and Legislative Analyst's Office. (2015). *Analysis of the impact of short-term rentals on housing*. San Francisco.

Camilleri, J., & Neuhofer, B. (2017). Value co-creation and co-destruction in the Airbnb sharing economy. *International Journal of Contemporary Hospitality Management, 29*(9), 2322–2340.

Cusumano, M. A. (2014). How traditional firms must compete in the sharing economy. *Communications of the ACM, 58*, 32–34.

Dahlander, L., & Magnusson, M. G. (2005). Relationships between open source software companies and communities: Observations from Nordic firms. *Research Policy, 34*, 481–493.

Denning, S. (2014). An economy of access is opening for business: Five strategies for success. *Strategic Leadership, 42*, 14–21.

Dubois, A., & Gadde, L. E. (2014). 'Systematic combining' – A decade later. *Journal of Business Research, 67*, 1277–1284. https://doi.org/10.1016/j.jbusres.2013.03.036.

Edelman, B., Luca, M., & Svirsky, D. (2017). Racial discrimination in the sharing economy: Evidence from a field experiment. *American Economic Journal: Applied Economics, 9*(2), 1–22.

Ert, E., Fleischer, A., & Magen, N. (2016). Trust and reputation in the sharing economy: The role of personal photos in Airbnb. *Tourism Management, 55*, 62–73.

Firnkorn, J., & Müller, M. (2011). What will be the environmental effects of new free-floating car-sharing systems? The case of Car2go in Ulm. *Ecological Economics, 70*(8), 1519–1528.

Flint, D. J., & Woodruff, R. B. (2001). The initiators of changes in customers' desired value: Results from a theory building study. *Industrial Marketing Management, 30*, 321–337.

Franke, N., & Shah, S. (2003). How communities support innovative activities: An exploration of assistance and sharing among end-users. *Research Policy, 32*, 157–178.

Funk, B., Bachmann, A., Becker, A., Buerckner, D., Hilker, M., Kock, F., Lehmann, M., & Tiburtius, P. (2011). Online peer-to-peer lending? A literature review. *Journal of Internet Banking and Commerce, 16*, 1–18.

Gale, B., & Wood, R. C. (1994). *Managing customer value: Creating quality and service that customers can see*. New York: Simon and Schuster.

Gallarza, M. G., Gil-Saura, I., & Holbrook, M. B. (2011). The value of value: Further excursions on the meaning and role of customer value. *Journal of Consumer Behavior, 10*, 179–191.

Gansky, L. (2010). *The mesh: Why the future of business is sharing*. New York: Penguin.

Garbarino, E., & Strahilevitz, M. (2004). Gender differences in the perceived risk of buying online and the effects of receiving a site recommendation. *Journal of Business Research, 57*, 768–775.

Guttentag, D. (2013). Airbnb: Disruptive innovation and the rise of an informal tourism accommodation sector. *Current Issues in Tourism, 18*, 1–26.

Guttentag, D., Smith, S., Potwarka, L., & Havitz, M. (2017). Why tourists choose Airbnb: A motivation-based segmentation study. *Journal of Travel Research, 56*, 1–18.

Hamari, J., Sjöklint, M., & Ukkonen, A. (2016). The sharing economy: Why people participate in collaborative consumption. *Journal of the Association for Information Science and Technology, 67*, 2047–2059.

Holbrook, M. B. (Ed.). (1999). *Consumer value: A framework for analysis and research*. London: Routledge.

John, N. A. (2012). Sharing and web 2.0: The emergence of a keyword. *New Media and Society, 15*, 167–182.

John, N. A. (2013). The social logics of sharing. *The Communication Review, 16*, 113–131.

Katzev, R. (2003). Car sharing: A new approach to urban transportation problems. *Analyses of Social Issues and Public Policy, 3*, 65–86.

Kneese, T., Rosenblat, A., & boyd, d. (2014). *Understanding fair labor practices in a networked age*. Data & Society Research Institute. https://www.datasociety.net/pubs/fow/FairLabor.pdf. Accessed 1 Apr 2017.

Lahti, V.-M., & Selosmaa, J. (2013). *A fair share: Towards a new collaborative economy*. Helsinki: Atena.

Lamberton, C. P., & Rose, R. L. (2012). When is ours better than mine? A framework for understanding and altering participation in commercial sharing systems. *Journal of Marketing, 76*, 109–125.

Lerner, J., & Tirole, J. (2002). Some simple economics of open source. *Journal of Industrial Economics, 50*, 197–234.

Malhotra, A., & Van Alstyne, M. (2014). The dark side of the sharing economy and how to lighten it. *Communications of the ACM, 57*, 24–27.
Matzler, K., Veider, V., & Kathan, W. (2015). Adapting to the sharing economy. *MIT Sloan Management Review, 56*, 71–77.
Milanova, V., & Maas, P. (2017). Sharing intangibles: Uncovering individual motives for engagement in a sharing service setting. *Journal of Business Research, 75*, 159–171.
Moriarty, R. T. (1983). *Industrial buying behavior: Concepts, issues, and applications*. Free Press. Lexington, MA: D.C. Heath & Company.
Owyang, J. (2016). *Collaborative economy honeycomb*. http://www.web-strategist.com/blog/2016/03/10/honeycomb-3-0-the-collaborative-economy-market-expansion-sxsw/. Accessed 30 Mar 2017.
Patton, M. Q. (1990). *Qualitative evaluation and research methods*. California: SAGE Publications.
Piscicelli, L., Cooper, T., & Fisher, T. (2015). The role of values in collaborative consumption: Insights from a product–service system for lending and borrowing in the UK. *Journal of Cleaner Production, 97*, 21–29.
PwC. (2014). *UK sharing economy generates 9 billion by 2025*. http://www.consultancy.uk/news/1085/pwc-uk-sharing-economy-generates-9-billion-by-2025. Accessed 27 Apr 2017.
PwC. (2016). *Europe's five key sharing economy sectors could deliver €570 billion by 2025*. Retrieved May 5, 2017, https://press.pwc.com/News-releases/europe-s-five-key-sharing-economy-sectors-could-deliver--570- billion-by-2025/s/45858e92-e1a7-4466-a011-a7f6b9bb488f
Rintamäki, T., & Kirves, K. (2017). From perceptions to propositions: Profiling customer value across retail contexts. *Journal of Retailing and Consumer Services, 37*, 159–167.
Rintamäki, T., Kuusela, H., & Mitronen, L. (2007). Identifying competitive customer value propositions in retailing. *Managing Service Quality, 17*, 621–634.
Sheth, J. N., Newman, B. I., & Gross, B. L. (1991). Why we buy what we buy: A theory of consumption values. *Journal of Business Research, 22*, 159–170.
Smith, J., & Colgate, M. (2007). Customer value creation: A practical framework. *Journal of Marketing Theory and Practice, 15*, 7–23.
Smith, G. E., & Nagle, T. T. (2005). A question of value. *Marketing Management, 14*, 38–43.
Smolka, C., & Hienerth, C. (2014). The best of both worlds: Conceptualizing trade-offs between openness and closedness for sharing economy models. In *12th International Open and User Innovation Conference (2014)*, 70.
Stake, R. E. (2004). *Standards-based and responsive evaluation*. Thousands Oaks: SAGE.
Teubner, T. (2014). Thoughts on the sharing economy. *Proceedings of the International Conference on e-Commerce, 11*, 322–326.
Truffer, B. (2003). User-led innovation processes: The development of professional car sharing by environmentally concerned citizens. *Innovation: The European Journal of Social Science Research, 16*, 139–154.
Tussyadiah, I. P. (2015). An exploratory study on drivers and deterrents of collaborative consumption in travel. *Information and Communication Technologies in Tourism, 2015*, 817–830.
Tussyadiah, I. P., & Zach, F. (2017). Identifying salient attributes of peer-to-peer accommodation experience. *Journal of Travel & Tourism Marketing, 34*(5), 632–652.
Weitzman, M. L. (1986). Macroeconomic implications of profit sharing. In *NBER macroeconomics annual 1986* (Vol. 1, pp. 291–354). Cambridge, MA: MIT Press.
Woodruff, R. B. (1997). Customer value: The next source for competitive advantage. *Journal of the Academy of Marketing Science, 25*(2), 139.
Yadav, M. S. (2010). The decline of conceptual articles and implications for knowledge development. *Journal of Marketing, 74*, 1–19.
Yin, R. K. (1994). *Case study research: Design and methods*. Beverly Hills. Thousand Oaks: Sage Publications.
Yin, R. K. (1999). Enhancing the quality of case studies in health services research. *Health Services Research, 34*, 1209–1224.
Zeithaml, V. A. (1988). Consumer perceptions of price, quality, and value: A means-end model and synthesis of evidence. *Journal of Marketing, 52*, 2–22.

Back to the Future: A Revelation of Conventional Platform Preference of Digital Creative Ecosystem Entities in Bandung

Santi Novani, Cici Cintyawati, and Lidia Mayangsari

Abstract This chapter explores the application of service science to enhance the development of digital creative industry in Bandung, Indonesia. The digital creative industry is one of the creative industry sectors that will grow rapidly in the near future. Our objectives are to analyze the current practices of value co-creation (i.e., collaboration) in the digital creative industry with the theoretical frameworks of service science and to contribute to the value orchestration platform development strategy of the case ecosystem. We applied exploratory research as a methodology. At the first stage, we identify the critical issues of the digital creative industrial cluster. Based on the findings, we developed a value co-creation process model. Secondly, we developed a value orchestration model by evaluating the stakeholder's value co-creation strategies. The results show that stakeholders in the digital creative industry in Bandung prefer a physical platform like a focus group discussion (FGD) over digital platforms. Face-to-face discussion is perceived to result in better business decisions. This finding is in line with the principle of "musyawarah mufakat," which has become a defining cultural characteristic of Indonesian society.

Keywords Service ecosystem innovation · Value orchestration platform · Digital creative industry

S. Novani (✉) · C. Cintyawati · L. Mayangsari
School of Business and Management, Institut Teknologi Bandung, Bandung, Indonesia
e-mail: snovani@sbm-itb.ac.id; cici.cintyawati@sbm-itb.ac.id; lidia.mayangsari@sbm-itb.ac.id

© Springer Nature Singapore Pte Ltd. 2018

A. Smedlund et al. (eds.), *Collaborative Value Co-creation in the Platform Economy*, Translational Systems Sciences 11,
https://doi.org/10.1007/978-981-10-8956-5_13

Introduction

A digital creative industry that relies on technology, creativity, and intellectual property as a major asset has developed quite rapidly in recent years (AIMIA 2005; Sneha 2016; Leung and Bentley 2017). The digital creative industry has a different sub-sector including computers and online games, digital music, visual effects and animation, and software development, as well as e-health, e-tourism, e-commerce, and e-learning software (Proctor-Thomson 2013; AIMIA 2005; Leung and Bentley 2017).

Today, the growth rate of the creative industry in Indonesia is high; however, according to the data compiled by the Indonesian Central Statistics Agency about the Creative Economy (EKRAF), the contribution of the creative economy to the Indonesian GDP is quite low, just around 7% per year (Hermansyah 2016).This presents an opportunity and challenge for the digital creative industry to greatly expand in the near future. In Indonesia, the digital creative industry is predicted to continue to grow, along with the rapid growth in digital technology (Das et al. 2016). Digital technology is expected to have a positive impact on the economic development in Indonesia. The digital technology sector is projected to increase the annual economic income by generating almost USD 150 billion in 2025 (Das et al. 2016). This presents a great challenge and opportunity for developing countries like Indonesia to increase the development of the digital creative industry to generate economic benefits in the long run.

However, to enhance the development of digital creative industry in Indonesia, all stakeholders involved need to collaborate to create a healthy ecosystem. Collaboration has become an important issue today because it is perceived to contribute to innovativeness (Todeva and Knoke 2005). In a collaborative framework, each stakeholder must be able to work together (collaborate) in creating shared value (value co-creation). This concept is very similar to the concept of *gotong-royong* in Indonesia. *Gotong-royong* is a unique characteristic of Indonesia, which continues to be implemented until now. In this concept, people will collaborate to help each other in doing their work. The essence of this concept is collaboration.

In response to the collaboration goal in the ecosystem, it is important to create a platform as a media of communication and coordination among stakeholders. Research by Kijima et al. (2013) on the service ecosystem shows that in the service ecosystem, key players play a role in the platform that orchestrates and facilitates value co-creation by customers and providers. The platform is a strategy pursued by key players as a tool to invite stakeholders to participate in its platform. Afterward, the curation stage gives new meaning to products, information, or a service defined as the result of a collecting, selecting, analyzing, editing, and reexamining process. The platform is then used to facilitate the value co-creation process involving customers, providers, information, and technology based on the new meaning. The final destination of performing a value orchestration strategy is value co-creation that results from empowering customers and providers in a spiral process. Kijima, Rintamki, and Mitronen developed a theory that values an orchestration platform.

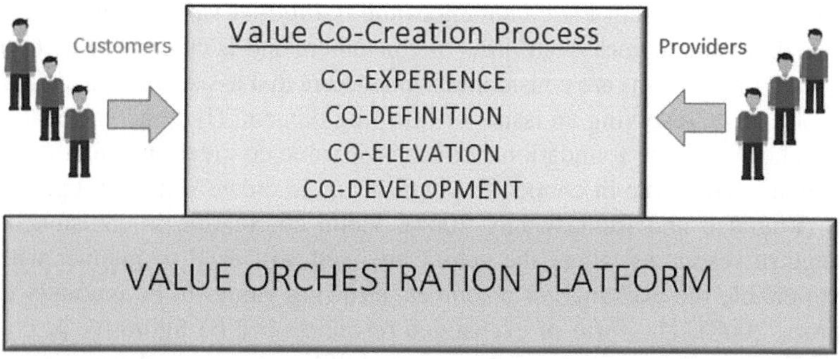

Fig. 1 Two layers of service systems

Figure 1 shows customers and providers encouraged to interact with each other and to co-create values often but without using information and communication technology (ICT).

A value orchestration strategy is a method used by an organization to create a healthy ecosystem with platform offered as key factor to succeed. There are three strategies proposed to orchestrate the platform, i.e., involvement, curation, and empowerment strategy. This paper analyzes a value orchestration platform developed in the digital creative industry. We will discuss the current practices of value co-creation (collaboration) in the digital creative industry using the framework of the service science and strategies used in creating a healthy ecosystem by using value orchestration strategies.

Theoretical Background

To gain a deeper understanding of this research area, we reviewed literature related to the concept of service science and service ecosystem innovation. Then, to analyze the results of this research, service science was used as the major framework for this research.

Value Co-creation in Service Ecosystem

The service ecosystem is defined as a relatively self-contained, self-adjusting system of resource-integrating actors, shared institutional logics, and mutual value creation through service exchange (Lusch and Vargo 2014). It is combined with what is thought of as the external environment as part of resources that are integrated into the entire value co-creation process and the role of networks as mediators of value co-creation because they enable access to resources (Akaka et al. 2012).

The service-dominant logic challenges the traditional value creation logic as it suggests that value comes from firms to consumers and is co-created by both consumers and firms. Firms are considered contributors that assist consumers in achieving an objective, resolving an issue, or fulfilling demand (Bettencourt et al. 2014). On the other hand, the foundational concept of value co-creation is that consumers will play an active role in cooperating with firms to create value through different stages (Prahalad and Ramaswamy 2004). Value co-creation is envisioned as an exchange of resources where the actors involved will need to interact with each other to enable the exchange of resources, allowing values to be mutually created (Gronroos 2008). The form of exchanged resources can be figurative or concrete. Essentially, this theory suggests that actors fulfill their needs through a resource exchange in which actors obtain their demanded resources from others using social interactions between the parties.

Kijima et al. (2013) reported that there are four phases of value co-creation process: co-experience, co-definition, co-elevation, and co-development. In the co-experience process, actors may need to share a model to co-define a reciprocal understanding regarding the problem that is to be defined. Through the interaction of actors, the capabilities and expectations of others may be identified so that they share and co-define a similar model (Galbrun and Kijima 2009). Co-elevation is a spiral-up process of consumer expectations and the capabilities of providers. Greater value and quality flow from high consumer expectations. Finally, co-development emphasizes the co-innovation originating from collaboration among entities (Novani et al. 2015).

Value Orchestration Platform Strategy

In this study, we will focus on examining the concept of a value orchestration platform. A value platform orchestration is a method used to involve all stakeholders in the ecosystem (Perks et al. 2017). This platform can be a media of interactions (Kijima et al. 2013) to facilitate the collaboration process between providers and customers (Novani et al. 2014). Meanwhile, the role of orchestration is perceived to create and manage the networks in the ecosystem (Perks et al. 2017).

To co-create value among stakeholders, in this case, "customers" and "providers" often require greater effort to help them understand each other so that the platform can be more effective. It is important to create mutual understanding among stakeholders in the ecosystem. Therefore, to develop a value orchestration platform, there are three strategies proposed by Kijima et al. (2013), which are involvement, curation, and empowerment strategies (Novani et al. 2015).

1. Involvement Strategy

Involvement is an initial strategy used to attract and involve people to the platform. To reach the objective of this strategy, we can use a cycle process called sympathize, identify, participate, and share and spread (SIPS). This process is used to generate interest among stakeholders to join the platform (Kijima et al. 2013). People will be connected by using a platform, whether it is a real platform (face-to-face discussion or events) or virtual platform like using ICT (social media, website, etc.). It needs to earn "sympathy" from each stakeholder so that they can join the platform and become interested in collaborating with other stakeholders.

2. Curation Strategy

   The curation strategy is the next stage used after the stakeholders get involved in the platform. This strategy relates to the way of analyzing the current problems in the digital creative industry and taking steps to overcome them.

3. Empowerment Strategy

   In this strategy, we will focus on how to make the platform empower the stakeholders so that they can be motivated to interact and collaborate with each other. Each stakeholder is empowered by increasing their aspiration (from the customer side) and capability level (from the provider side).

Method

This study approached the co-creation platform preferences qualitatively while focusing on understanding the nature of research rather than the quantity of observed characteristics. Social reality is a human creation. Thus, this study interprets and contextualizes meanings that emerge from people's beliefs and practices. A qualitative multi-case study serves as the methodological framework where it creates a holistic view, bounded system, and elaborative case, through in-depth data collection of several related data sources (Creswell 2012). The knowledge from data collected provides a better understanding of the collaboration platform preference. As taught by Yin (2014), a case study in general is useful for examining contemporary cases without manipulation of behaviors or variables.

In this study, the phenomenon examined is a collaborative platform preference of digital creative ecosystem entities in the cities of Bandung and Cimahi. There are two cases involved: Institute for Innovation and Entrepreneurship Development ITB (LPIK ITB) and Cimahi Techno Park (CTP). This study focused on three subsectors of the digital creative industry (animation, video, and application and game development) to gain a better understanding of the use of the platform in enhancing the digital creative industry in Bandung and Cimahi. This study used semi-structured interviews, where it relies on a general interview protocol as a guide. Then the interviewers may address other aspects and give opportunities to the participants to say things that lie outside of the interview guidelines, and the interviewer could probe deeper into an area of interest (Carpenter and Lertpratchya 2016, p. 452). This paradigm is more appropriate to be used to reach the objective of this study. A "deep and experiential understanding," "thick descriptions," and "multiple realities" can be obtained from a qualitative case study approach (Grybovych 2012).

Case Selection

Bandung and Cimahi are two neighboring cities in West Java Province in Indonesia. These cities are known as a hub where innovation, creativity, and entrepreneurship flourish. The government has many programs to improve the quality of life of citizens and to foster economic development in these cities, whose future identities include a reputation for digital creativity. The cities are rich business incubators, but this study limits its scope to two digital creative incubators: the Institute for Innovation and Entrepreneurship Development (LPIK ITB) and Cimahi Techno Park (CTP). These three incubators were selected because they are focused on the incubation program for the digital creative start-ups. This study assumes that by focusing on these two case studies, a deeper understanding of the real condition of the digital creative industry can be achieved.

1. *The Institute for Innovation and Entrepreneurship Development ITB (LPIK ITB), Bandung*

   LPIK ITB is an incubator set up by ITB to help techno-based start-ups resolve problems faced in Indonesia by developing business innovation. LPIK was selected because in this incubator, there are many digital start-ups, especially start-ups that create software and animation. Also, LPIK is one of the incubators in Bandung that aims to enhance the development of innovation in the business ecosystem.
2. *Cimahi Techno Park (CTP)*

   CTP is a particular area constructed by the local government to develop technology-based economy industries. CTP was inaugurated in March 2017 and aims to support the development of innovative technology-based businesses. The techno park is an example of the concept of a technology-based area development, which combines a science and technology center to support creative economy development. CTP was selected because it aims to help digital start-ups, especially the start-ups in the field of animation and telematics.
3. *The Institute for Innovation and Entrepreneurship Development ITB (LPIK ITB), Bandung*

   Institute for Innovation and Entrepreneurship Development ITB (LPIK ITB) is an incubator set up by ITB to help techno-based start-ups resolve problems faced in Indonesia by developing business innovation. LPIK was selected because in this incubator, there are many digital start-ups, especially start-ups that create software and animation. Also, LPIK is one of the incubators in Bandung that aims to enhance the development of innovation in the business ecosystem.

4. *Cimahi Techno Park (CTP)*

   Cimahi Techno Park is a unique site constructed by Cimahi's government to develop technology-based economy industries. CTP was inaugurated in March 2017. CTP aims to support the development of the innovative technology-based business. The techno park building reflects the concept of a technology-based area, by combining a science and technology center to support creative economy development. CTP was selected because it aims to help digital start-ups, especially the start-ups in the field of animation and telematics.

Data Collection

Since this study aims to analyze the situation and condition of the digital creative ecosystem, where only limited research has been conducted, this research is carried out as exploratory research with a qualitative approach. Following the guidelines of doing exploratory research, we did a search for and review of appropriate literature, conducting in-depth interviews and focus group discussions. Since the stakeholders in the digital creative industry are digital start-ups, government, academia/university, incubators, associations, large enterprises or financial institutions, and customers, we only focused on choosing key people from academia, government, association, digital start-ups, and incubators as our interviewees. We did not choose all possible stakeholders because these five stakeholders were considered acceptable to provide sufficient information to develop a strong model and platform.

In selecting the interviewees, a judgmental sampling strategy was used. This strategy allowed us to choose the samples based on our judgment. In this case, we choose interviewees based on their potential contribution to the digital creative ecosystem but still relevant to fulfill the research objectives of this study.

Ten interviewees were chosen for this research. They included two individuals from government (Bandung and Cimahi), two individuals from incubators (CTP and LPIK), two individuals from academia (lecturers in ITB), two members from a community/association (CCA), and two start-ups that develop animation and software and also participate in an incubation program in LPIK and CTP. These interviewees play key roles in their organization, so their opinion was considered potentially important contributions in developing digital creative industry in Bandung and Cimahi.

The interview process averaged between 30 min and 1 h. The interviewees were asked questions about their role in developing the digital creative industry, problems or issues faced by the digital creative industry in Cimahi and Bandung, and their opinion of how to solve these issues or problems, especially related to the platform needed to enhance the coordination and collaboration of one stakeholder with other stakeholders. The interview process was conducted in Bahasa Indonesia to help them better understand the purpose of the interview and to avoid miscommunication so they can give clear opinions to the interviewer. The interview was conducted

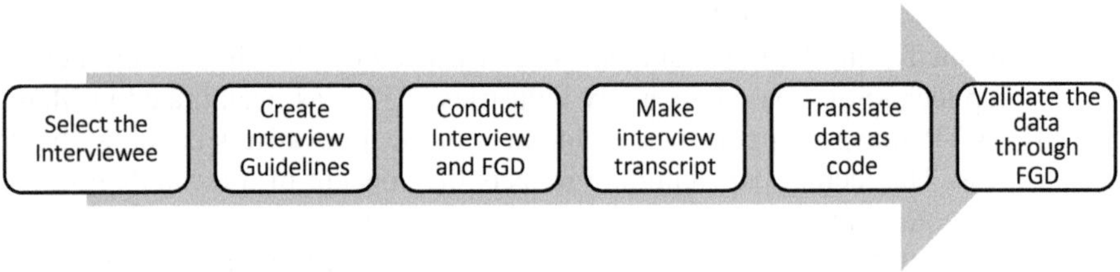

Fig. 2 Data collection process

using a one-to-one interview to gather different points of view and information from each informant. In addition, before we started the interview, we also asked for permission from each interviewee/informant to record the interview process. This recording was used to help us codded all of important information from the interviewees.

After conducting the interview, we then made transcripts of the interviews, which were then coded to make it easier to analyze the data collected. The data were collected from the stakeholders, which was then analyzed to draw conclusions and make recommendations. From the preliminary studies, one particular issue was selected to be the target of a more focused research question, which was to study the business ecosystem of digital creative industry, especially the existence of a value orchestration platform to increase the health of digital creative ecosystem. From the data collected, we could create a platform model that is more suitable to be used in the digital creative ecosystem.

In addition to the interview, we also conducted a focus group discussion to gather more points of view about the problems and to gather more insight about the solution needed to enhance the development of the digital creative industry in Bandung and Cimahi. This focus group discussion (FGD) was conducted twice. The first FGD on 12 September 2017 was conducted to gather information about the issues faced, and the second FGD on 7 November 2017 was to validate the solution proposed by the researchers. The data collected from the FGD also was also gathered in a transcript and coded.

The figure above explains the stages used to collect the data for this research (Fig. 2).

Data Analysis

The aims of this study were (1) to analyze the current practices of value co-creation (collaboration) in the digital creative industry using the framework of service science, (2) to contribute to the value orchestration platform development strategy of digital creative industry to be adapted by the stakeholders involved in the ecosystem, and (3) to contribute to the literature on this topic. These objectives drove the focus of this study to enhance the development of the digital creative industry in Indonesia, especially in Bandung.

To reach the aforementioned objectives, there are three stages used to analyze the data of this study. The first stage is case analysis to collect deep information about the application of value co-creation (collaboration) application in the digital creative industry. We also identified and described the digital creative ecosystem and interaction model among stakeholders in the ecosystem. Also, we identified the role of each stakeholder in the ecosystem to understand their potential and involvement in creating a healthy ecosystem. A semi-structured interview was conducted to gain more insight or opinion from each stakeholder about the current condition of the digital creative industry. A triangulation process through focus group discussion was conducted to validate the information obtained from each stakeholder.

In the second stage, a cross-case analysis was conducted (Yin 2014 quoted in Perks et al. 2017). In this stage, a focus group discussion was conducted to identify and analyze the most appropriate platform to be used by the stakeholders in the ecosystem. Then, we focused on the identification of the strategy used in developing a value orchestration platform in the digital creative industry.

In the second stage, we could uncover new strategies to be used in creating a healthy digital creative ecosystem. The last stage then was conducted to validate these strategies by getting feedback from the stakeholders. Each stakeholder gave their opinion about these new strategies whether these strategies were appropriate for them or not. So, the solution derived from this study can be applied to enhance the development of the digital creative industry in Indonesia, especially in Bandung.

Findings

Digital Creative Ecosystem

A digital creative ecosystem is a place where all of stakeholders are involved in enhancing the digital creative industry development. Based on the interview results, at least seven stakeholders are involved in the digital creative ecosystem, which include digital start-ups, university, government, association/community, customer, industry, and incubator. From the interview results, we tried to outline the ecosystem of digital creative industry in Bandung, as shown in the figure below (Fig. 3).

1. University

   The university plays a role as a major researcher that helps start-ups develop their idea or product and facilitate some business education for the start-ups. Also, the university can help start-ups by providing talent needed by start-ups.
2. Government

   The government has a role as facilitator and policymaker for the digital creative industry in Indonesia. For example, Indonesia's government created an agency called the Creative Economy Agency (BEKRAF) to assist the president in formulating, assigning, coordinating, and synchronizing policies in the creative economy in Indonesia. The major role of the government in the ecosystem

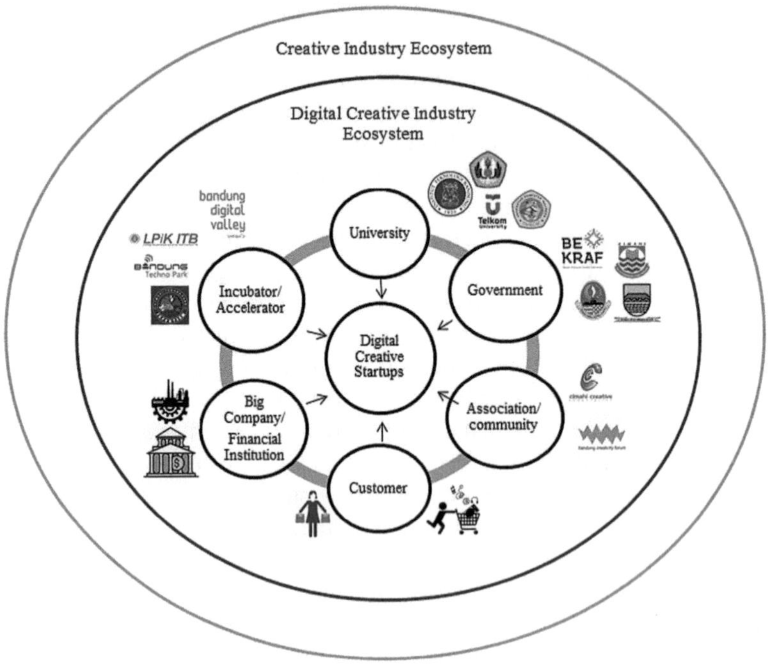

Fig. 3 Entities of the digital creative industry in Bandung

is as a policymaker, not as investors. The government will create some appropriate policies to help digital start-ups in developing their business.

> The Government can only help digital start-ups by giving some facilities like a techno-park building and other infrastructure. We didn't have enough in our budget to give money to all digital start-ups. We just created some policy that could help the start-ups development. (GVN025)

Based on the interview results, the role of the government actually is not to provide money to digital start-ups but to help the start-ups by creating policies that facilitate start-ups in developing their business. In addition, the government has a role in providing some facilities needed by start-ups.

3. Business Incubator or Accelerator

Business incubator has a role in helping start-ups in developing their business such as providing funds for start-ups and helping in increasing the skill of human resources in start-ups by creating an incubation or acceleration program.

> We provide incubation programs so start-ups can learn how to manage and develop business. After this, we give funds to any selected start-ups to help them develop their business. We also often hold activities that aim to bring together start-ups with investors. (INCB, 020)

She said that the role of incubators is to provide an incubation program for the digital start-ups. The incubators or accelerators will provide the mentor or coach to give business knowledge to the start-ups. The incubators also can give some funds to the selected start-ups so that they can use these funds to develop their business.

4. Association/Community

   Associations have a role as a meeting place for people who concerned about the development of the digital creative industries in Indonesia. An association can be a good media for sharing information and ideas or conducting some events in the digital creative industry.

> We have conducted many events aim to attract people. We usually conduct an annual event like the Bandung International Digital Arts Festival to exhibit many digital art collections around the world. (ACC008)

   He said that their role as a creative association is through facilitating and conducting some events so that many stakeholders will gather in one place to discuss the issues faced today. In addition, these events also become a platform for digital start-ups in showing or even marketing their products to consumers.

5. Big Company and Financial Institution

   The role of a big company and financial institution is to help the digital start-up by providing funds to develop and enhance their business. A big company also can be a partner of start-ups in developing their products.

6. Digital Start-Ups

   Digital start-ups are small companies that focus on producing digital technology-based products or services. The role of digital start-ups in the eco-system is as a company that makes innovative and creative digital products or services.

   In this research, the digital start-ups become the major focus of the ecosystem because based on the interview results, the most important thing is to create a healthy ecosystem by increasing the quantity and quality of digital start-ups.

> ...to make a healthy digital creative ecosystem, it is important to start by enhancing the quality, not only the number of digital start-ups. (ASS027)

   It believed that focusing on increasing the quality and quantity of digital start-ups is more than focusing on how to make more customers. This is because, if there is no innovation from start-ups due to the low quality of start-ups, then the number of start-ups will decrease significantly, and the ecosystem will be unbalanced. He thought attracting customers could be the next step to take after start-up businesses are well-developed. This is because the consumer of the digital start-ups is not only a personal customer but also an organizational customer (the government, the universities, or the investors). Also, from the S-D logic perspective, the customer is perceived as a value co-creator, where providers can collaborate with the customer to co-create value. But, in this case, although a customer appears as a stakeholder in the digital creative ecosystem, we found that they still have no role in the current ecosystem. The customer is assumed to be the value consumer only; thus, customers do not seem to have any participation in the ecosystem.

Based on the depth interview results, each stakeholder has different potential and involvement, which is high, medium, or low. The table below explains the potential and involvement of each stakeholder in the digital creative ecosystem in Bandung (Table 1).

The table below shows that in the digital creative ecosystem, there is a full set of roles, but many parties believe that the involvement of each stakeholder is not optimum. It should further be questioned, whether there is any reason that makes the involvement of each stakeholder not optimum and whether it is caused by the low coordination or different expectations among stakeholders.

Interaction Process Among Stakeholders

In addition to identifying the digital creative ecosystem model, we then tried to determine and analyze the interactions that occur between the parties involved in the ecosystem. This interaction process can help the stakeholders in understanding the role of each other in the ecosystem. In-depth interviews and observations were conducted to draw that interaction. We tried to explain the interaction happening in the ecosystem through the interaction model below (Fig. 4).

Table 1 Categorization of stakeholders' role in the ecosystem

	Characteristic	
Type of stakeholder	Potential	Involvement
Digital start-ups	High	High
University/academia	Medium	Medium
Incubator/accelerator	High	High
Association/community	Medium	Medium
Government	High	Medium

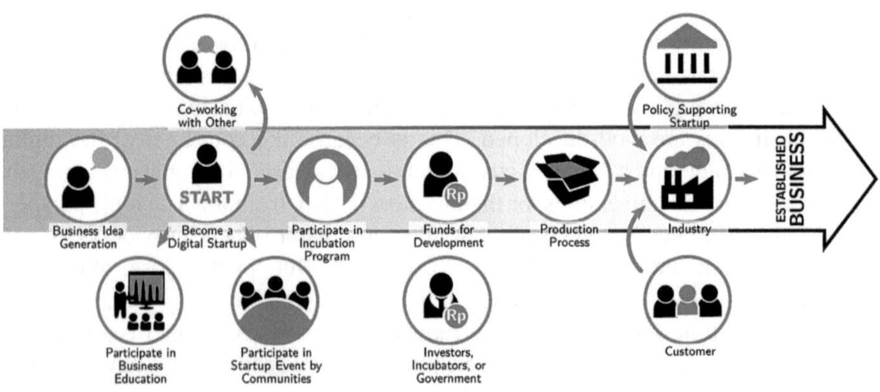

Fig. 4 Interaction model among stakeholders in the digital ecosystem

Since each stakeholder has a different role in the ecosystem, collaboration and innovation in the service ecosystem are necessary to create a healthy ecosystem. But unfortunately, the results of the explorative study show that the collaboration among stakeholders in the digital creative industry is still uncommon. This can be caused by a lack of communication and coordination among stakeholders. Therefore, the role of the platform in the digital creative industry is very critical.

> ...in fact, there has been collaboration among stakeholders in the ecosystem. However, the collaboration is still very low, whereas the collaboration is necessary to help us in achieving our goals. (ACD007)

Based on the interview results above, collaboration among stakeholders has been happening but remains weak. Furthermore, they also consider the fact that stakeholders have limited resources and capabilities, so they need to integrate and collaborate with others to innovate and deliver better products or services through an orchestrator or independently.

> We can create our product by ourselves. We don't need to collaborate with other parties. (DSP012)

But, there are some stakeholders who believe they can achieve their goals without having to collaborate with other stakeholders. This means that the mutual understanding between them is still undeveloped. This can be the result of a lack of communication and interaction among them. The collaboration issue then became a major issue to be solved in this research.

Value Orchestration Platform

After conducting ten interviewees (from academia, government, incubator, digital start-ups, and association), almost all stakeholders believed that they needed a platform to help them better communicate and coordinate with each other in developing a digital creative industry. The platform is viewed as an effective media in co-creating the value among stakeholders by involving stakeholders, information, and technology in one place.

> Platforms can be an effective media for improving coordination and communication among stakeholders. We believe that the platform is necessary to make it easier for us to interact and collaborate with each other so that there is no misunderstanding in the fulfillment of our roles..... (ASS056)

They assumed that a platform is needed to make it easier to communicate with each other. This could be the reason why a platform is important in enhancing collaboration among the stakeholders in the ecosystem. If effective communication and coordination can be achieved through a platform, then this can facilitate the process of value co-creation within the ecosystem.

Before analyzing a platform, we will focus on some issues faced by the stakeholders in the digital creative industry derived from a qualitative approach. These issues include difficulties in getting funding, a lack of trust among the stakeholders, and lack of human resources (talent and expertise) to develop innovation.

> There are several issues faced by us as the digital start-ups like lack of funds, lack of talent.... (DSP034)

> We do not collaborate with other parties because there is no guarantee they can be trusted to help us in developing our business. (DSP050)

One issue faced by the digital start-ups is a lack of funding; this means that they have challenges in collecting funds to develop their business. This issue can be solved by collaborating with the incubators, large enterprises, or financial institutions to obtain funding to run the business. Another issue is a lack of talent and expertise. Some digital start-ups have trouble finding appropriate human resources and trained staff to run their business. This is because they need to hire people with necessary capabilities and capacities like people with skills in developing creative digital products. This problem can be solved by collaborating with universities that can provide appropriate talent needed by the start-up. Next, a lack of trust becomes an issue faced by most of the stakeholders. This issue caused by trust among the stakeholders is still limited. Many of them believed that other parties could not be trusted because each was focused on achieving their goals. This can be caused by a lack of communication and interaction among them. These collaboration issues can be a good reason as to why a platform is important in solving these problems.

In realizing the value co-creation process, there is a strategy used to increase collaboration among stakeholders called a value orchestration platform. A value orchestration platform encourages stakeholders to interact with each other to create new values using ICT. In the value orchestration platform model, we will not only focus on the strategy to increase the number of digital start-ups but also the quality of start-ups, so the left side of this platform is digital start-ups itself. Below is a model of the two-sided platform to develop the digital creative industry in Bandung (Fig. 5).

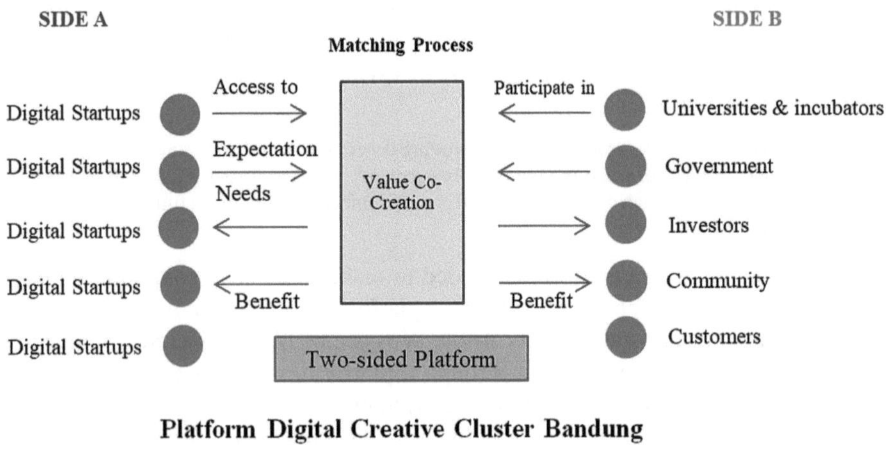

Fig. 5 Model of two-sided platform in the digital creative industry

Table 2 Stakeholder implementation in value orchestration strategy

Stakeholders	Involvement	Curation	Empowerment
University/ academia	Introduce the vision of "increasing the quality and quantity of digital start-ups"	Willingness to conduct research with digital start-ups	Initiate collaboration with others
	Establishing the real platform: FGD and seminars to introduce the vision of "increasing the quality and quantity of digital start-ups"	Willingness to facilitate a talent development program with digital start-ups	Platform is embedded
		Establishing education program on how to make digital products (animation, software, etc.)	Commit to its role
		Intention to collaborate with SMEs in making digital products	Co-production is embedded
Community	Establishing platform: Baros International Animation Festival (BIAF), *Bandung* International *Digital* Arts Festival (BIDAF), IXPO, etc.	Community informing and advertising new trend in the digital creative industry	Orchestrate stakeholders in harmony
Government	Introduced to the vision of "increasing the quality and quantity of digital start-ups"	Initiate 1,000 Digital Startups National Movement program	Regulator function (produce regulation, monitor, and evaluate)
	Approach for media to announce the platform	Convincing stakeholders that digital creative market is national and overseas customer	Shared institution is embedded
	Road show to region to inform the platform	Encouraging stakeholders that digital creative products are one of the multinational products	–
Incubators/ accelerator	Introduced to the vision of "increasing the quality and quantity of digital start-ups"	Facilitating incubation programs (funding and start-up development)	Commit to its role
		Intention to join the platform	Resource/skill sharing

A value orchestration strategy is a strategy used to create a healthy ecosystem by using a platform as the key factors to succeed. The strategy begins with an involvement strategy. An involvement strategy aims to invite stakeholders on board to join the platform. The next strategy is the curation strategy and then the empowerment strategy.

After conducting some in-depth interviews, the process of digital creative stakeholders in a value orchestration strategy is described in Table 2.

1. Involvement Strategy

In increasing the spirit of co-creation, it is necessary to have direct involvement from each stakeholder involved in the digital creative ecosystem. Communication is the first step used to increase the understanding of each stakeholder. This process is the first step to create co-experience. If the co-experience is created, then it is expected that concerned stakeholders can understand each other's wishes, expectations, or capabilities so that no one party feels disadvantaged (co-definition).

The concept of value co-creation among stakeholders can be created if each stakeholder can benefit from this co-creation process. For example, the digital start-ups increase their expectations of the facility received, and the government improves the facility provided.

Therefore, each stakeholder needs to know and communicate the benefits they want to attain when they are involved in the digital creative ecosystem. For example, a major issue in the digital creative industry is increasing the quantity and quality of digital start-ups. A university can act as a platform provider and then introduce the vision to other stakeholders. Here each stakeholder needs to know the desire of other stakeholders to create a healthy digital creative ecosystem. Therefore, the platform becomes a media of interaction among stakeholders.

The platform can facilitate communication and coordination among stakeholders (especially between government, digital start-ups, incubator, university, and association). The platform is used to realize the involvement strategy, where the government involves the digital start-ups, incubator, university, and incubators in making some policies for the digital creative industrial cluster. The platform can be either physical or virtual. Physical platforms can be a seminar, focus group discussion (FGD), and other activities or even a place that can facilitate interaction among stakeholders.

Based on the results of exploratory research, the stakeholders in the digital creative industry prefer a physical platform (face-to-face platform) to a virtual platform (social media).

> A Face-to-face platform is considered more effective to be applied in the digital creative industry than face-to-screen because in face-to-face communication, an agreement will be easier to reach… (ASS078)

> A virtual platform will be difficult to use by the government because not all government staff are interested in using virtual platform like a website or other social media to discuss something. (GVN015)

Almost all stakeholders prefer a face-to-face platform (real platform). This is because, with face-to-face communication, they will be more able to express their needs and expectation. Through activities like seminars, forums, or focus group discussion (FGD), the government can interact with the digital start-ups and other stakeholders to directly hear their opinions and aspirations. Such forums are perceived to be more effective because they can facilitate stakeholders in reaching an agreement.

> We can easily discuss the problems faced by the digital creative industry in the forum like this (focus group discussion). (GVN009)

This is in line with a principle held firmly by the Indonesian people, in which decision-making usually is made based on the principle of deliberation or referred to as the principle of *musyawarah mufakat.* To make this process more effective, feedback from each stakeholder is important. Feedback can be used to increase the service of each stakeholder in the ecosystem. For example, while conducting a focus group discussion, many digital start-ups complained about the difficulty of the licensing processes in establishing a limited company (PT). This can be good insight to help the government simplify the licensing process.

Besides using FGD or other forums, there have been some events conducted to involve the digital start-ups and other stakeholders in the digital creative industry. There have been some events conducted by the community before, like the Baros International Animation Festival (BIAF), Bandung International Digital Arts Festival (BIDAF), IXPO (Informatics Explosion), and other events. These events can be a platform to increase the public interest toward the digital creative industry. The government, incubators, universities, or communities can be a facilitator to create some digital creative events that can attract people to get involved in the digital creative industrial cluster.

The involvement strategy is part of the sympathize, identify, participate, and share and spread (SIPS) cycles aimed at fostering public interest in participating in the platform (Dentsu 2011). The platform is used to make people curious and concerned so that people have an interest to know more about the issues faced by the digital creative industrial cluster. This aims to increase the level of public participation in the developing digital creative industry. If people are willing to participate, they tend to spread information to others so that more people will be involved. This can create a spirit of value co-creation among stakeholders. The figure below explains the process used to reach the co-experience and co-definition process (Fig. 6).

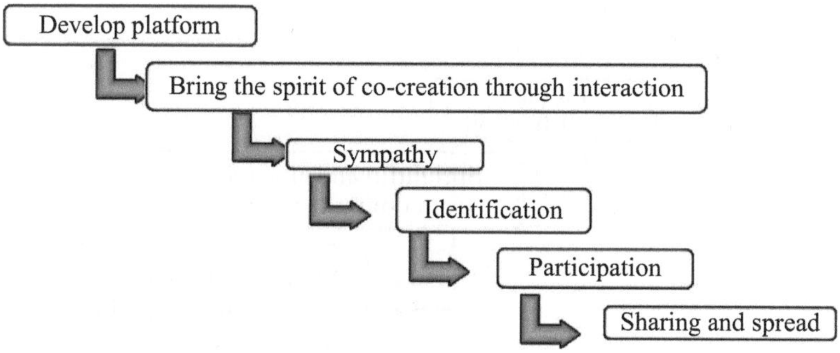

Fig. 6 Process of co-experience and co-definition (Dentsu 2011)

2. Curation Strategy

Curation strategy relates to a way of analyzing the current condition of the digital creative industry and taking steps to overcome it. This strategy is used to reach co-elevation and co-development. This strategy is related to proactive steps that need to be taken by the government in overcoming problems in the industry. For example, to increase the number and quality of digital start-ups, the government created the 1,000 Digital Startups National Movement.

> BEKRAF has created a program called the 1,000-Digital Startups National Movement. This program aims to give business knowledge to the digital start-ups so they can manage and develop their business. (INC024)

She said that currently, the curation strategy developed by the government is to create the 1000 Digital Startups National Movement. This program aims to increase the number of digital creative start-ups in Indonesia. By creating this program, the government expects that the number of digital creative start-ups will increase to enhance the development of a digital creative ecosystem in Indonesia.

Next, the university or academia has initiated joint research or co-production with digital start-ups and then helps them by providing high-capability talent for the start-ups. Incubators create an incubation program and initiate having others join the platform. The community has informed and advertised a new trend in the digital creative industry. Next, other parties became aware of the platform and intended to join and, further, were inviting others to adhere to the platform.

The involvement and the alertness of each stakeholder in overcoming the existing problems faced by the digital creative industry will make people more interested in participating (be motivated) and can invite others to participate in the ecosystem. Each stakeholder needs to have consistency in realizing every plan that was made so that it can increase the trust among stakeholders. In achieving the plan, each stakeholder needs to consider the benefit and impact of the action taken so that each stakeholder can gain some benefit from this co-creation process. Below are the co-elevation steps in co-creation (Fig. 7).

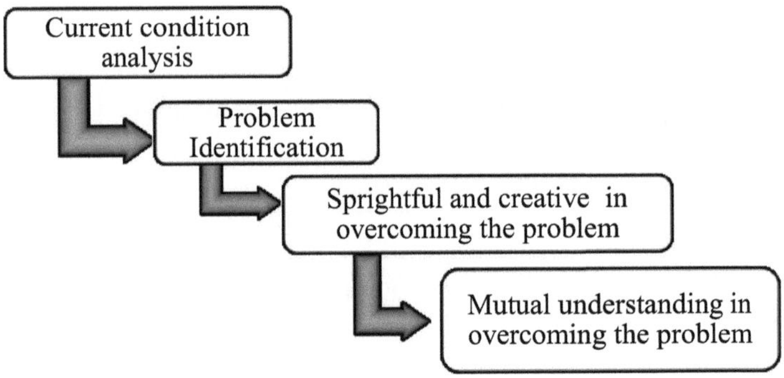

Fig. 7 The process of co-elevation

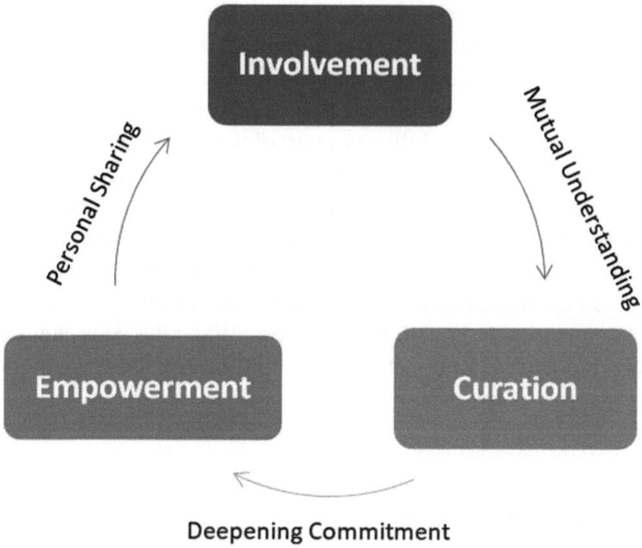

Fig. 8 Cyclic process of value orchestration strategy

3. Empowerment Strategy

In the empowerment strategy, each stakeholder uses the platform effectively to increase interactions among each other. Based on the interview and observation results, the empowerment strategy has not yet happened in the digital creative ecosystem. It can be concluded that value orchestration strategies in the digital creative ecosystem through FGD and event platforms have only achieved the involvement and curation stage. This can be due to the lack of effectiveness of FGD and other events aiming to increase the quality and quantity of the digital creative industry.

Learning from the digital creative ecosystem by using the lens of the service industry, we conclude that the process of the value orchestration strategy in the digital creative ecosystem should not stop at the empowerment stage. It should be a cyclic process as proposed in Fig. 8. Also, it is found that there is another stage needed before stepping into the involvement, curation, and empowerment stage. The figure above explains the cycle process needed to create a healthy digital creative ecosystem.

Before the involvement stage, the initiation stage is needed to help people open their minds about the platform. This initiation stage is referred to as personal sharing. The stakeholders are expected to share their personal desire and expectation to make them more interested in joining the platform. The involvement stage is then initiated by inviting the stakeholders to join the platform. In the involvement stage, the stakeholders are expected to have mutual understanding so they can continue the curation stage. Mutual understanding is needed to make the stakeholders choose a good solution by still considering the other stakeholders' needs. Next, in the curation stage, the orchestrator will propose a new meaning of products, services, or information (service) to stakeholders to produce better products, services, or information for the entire ecosystem. High commitment to fulfill their role in the ecosystem will be a sign that an ecosystem is ready for the next stage – empowerment. If each stakeholder has high commitment to be a part of the ecosystem, then they will consider their role and responsibility as a part of the ecosystem so the empowerment stage can be achieved. After achieving the empowerment stage, the next issues may appear so the process would be back to the involvement stage when the vision of the platform is achieved. The process does not stop at the end of empowerment process; instead, a new platform will be introduced to stakeholders as a new involvement stage strategy. In the digital creative ecosystem, the current vision of the platform is to increase the quantity and quality of start-ups. When the ecosystem has achieved the platform's objectives, a new platform should be created to keep the ecosystem healthy.

Discussion and Theoretical Contribution

Value co-creation can be created if each stakeholder is motivated to work together. The motivation can be generated from the interaction and communication in the platform, like motivation to get new knowledge, to interact with other stakeholders, and to create a partnership, or motivation because they feel involved (feel like they are an important part) in developing the digital creative industry, and others. The platform can serve as a media to effective communication and coordination so co-creation can be achieved.

The process of value co-creation in the digital creative industry is quite similar to other creative industries. But, since in this research we only focus on a start-up's development, the customer here is the digital start-ups themselves. Then, the provider may be the government, universities, incubators, or communities that help digital start-ups in developing their business.

Strategies used by the stakeholders in the value co-creation process are unique because almost all stakeholders prefer more to use a physical platform (FGD, forum, discussion, and events). This is because they believe it provides more effective communication if they have face-to-face interaction with the other stakeholders. This concept is consistent with the principle of *musyawarah mufakat.* This principle is still firmly held by most Indonesians in making decisions or solving problems. A face-to-face platform is perceived as the more effective platform to be used in the digital creative industry.

Limitations and Further Research

One limitation of this research is that it was conducted with a limited number of participants from Bandung. For future research, the scope of the research could be broader, for example, using an Indonesian case study as a whole. In addition, future research could be more specific or, for example, focusing on animation, software development, or the game sub-sector so that the exploration could be deeper using one or two specific sub-sectors of the digital creative industry.

References

Akaka, M. A., Vargo, S. L., & Lusch, R. F. (2012). An exploration of networks in value cocreation: A service-ecosystems view. *Review of Marketing Research, 9*, 13–50.

Australian Interactive Media Industry Association (AIMIA). (2005). *Digital content industry Roadmapping study*. Sydney: AIMIA.

Bettencourt, L. A., Lusch, R. F., & Vargo, S. L. (2014). A service lens on value creation: Marketing's role in achieving strategic advantage. *California Management Review, 57*(1), 44–66.

Carpenter, S., & Lertpratchya, A. P. (2016). A qualitative and quantitative study of social media communicators: An extension of role theory to digital media workers. *Journal of Broadcasting & Electronic Media, 60*(3), 448–464.

Creswell, J. (2012). *Educational research: planning, conducting, and evaluating quantitative and qualitative research* (4th ed.). Boston: Pearson.

Das, K., Gryseels, M., Sudhir, P., & Tan, K. T. (2016). *Unlocking Indonesia's digital opportunity* (McKinsey & Company Indonesia Report). Dentsu (2011) SIPS. www.dentsu.co.jp/sips/index.html

Dentsu. (2011). *SIPS*. www.dentsu.co.jp/sips/index.html. Accessed 18 May 2014.

Galbrun, J., & Kijima, K. (2009). Fostering innovation system of a firm with hierarchy theory: Narratives on emergent clinical solutions in healthcare. In *Proceedings of the 52nd Annual Meeting of the ISSS.*

Gronroos, C. (2008). Service logic revisited: Who creates value? And who co-creates? *European Business Review, 20*(4), 298–314.

Grybovych, O. (2012). Designing a qualitative multi-case research study to examine participatory community tourism planning practices. In K. F. Hyde, C. Ryan, & A. G. Woodside (Ed.), *Field guide to case study research in tourism, hospitality and leisure* (Advances in culture, tourism and hospitality research, Vol. 6, pp. 501–520). Bingley: Emerald Group Publishing Limited.

Hermansyah, A. (2016). Creative economic sector grows 62 percent in five years. *The Jakarta Post,* published in December 8, 2016. http://www.thejakartapost.com/news/2016/12/08/creative-economic-sector-grows-62-percent-in-five-years.html

Kijima, K., Rintamki, T., & Mitronen, L. (2013). *Value orchestration platform: Model and strategies* (pp. 13–16). Japan: In Annual conference of Japan society for management information.

Leung, L., & Bentley, N. (2017). Producing leisured laborers: Developing higher education courses for the digital creative industries. *The Journal of Arts Management, Law, and Society, 47*(2), 148–160.

Lusch, R. F., & Vargo, S. L. (2014). *Service-dominant logic: Premises, perspectives, possibilities.* Cambridge: Cambridge University Press.

Novani, S., Putro, U. S., & Hermawan, P. (2014). An application of soft system methodology in Batik industrial cluster solo by using service system science perspective. *Procedia- Social and Behavioral Sciences, 115*(1), 324–331.

Novani, S., Putro, U. S., & Hermawan, P. (2015). Value orchestration platform: Promoting tourism in batik industrial cluster Solo. *Procedia-Social and Behavioral Science, 169*, 207–216.
Perks, H., Kowalkowski, C., Witell, L., & Gustafsson, A. (2017). Network Orchestration for value platform development. *Industrial Marketing Management, 67*, 106–121.
Prahalad, C. K., & Ramaswamy, V. (2004). Co-creation experiences: The next practice in value creation. *Journal of Interactive Marketing, 18*(3), 5–14.
Proctor-Thomson, S. (2013). Gender disruptions in the digital industries? *Culture & Organization, 19*(2), 85–104.
Sneha, P. P. (2016). *Studying digital creative industries in India: Initial questions*. Centre for Internet & Society. Available in: http://cis-india.org/raw/studying-digital-creative-industries-in-india-initial-questions. Accessed 5 May 2017.
Todeva, E., & Knoke, D. (2005). Strategic alliances and models of collaboration. *Management Decision, 43*(1), 123–148.
Yin, R. K. (2014). *Case study research: Design and methods*. New Delhi: Sage Publications.

MIX
Papier aus verantwortungsvollen Quellen
Paper from responsible sources
FSC® C105338

If you have any concerns about our products,
you can contact us on
ProductSafety@springernature.com

In case Publisher is established outside the EU,
the EU authorized representative is:
Springer Nature Customer Service Center GmbH
Europaplatz 3, 69115 Heidelberg, Germany

Printed by Libri Plureos GmbH
in Hamburg, Germany